C000213544

POOR ROBIN'S PROPHECIES

POOR ROBIN's PROPHECIES

A curious Almanac,
and the
everyday mathematics of
Georgian Britain

Benjamin Wardhaugh

Great Clarendon Street, Oxford, OX2 6DP,
United Kingdom

Oxford University Press is a department of the University of Oxford.
It furthers the University's objective of excellence in research, scholarship,
and education by publishing worldwide. Oxford is a registered trade mark of
Oxford University Press in the UK and in certain other countries

© Benjamin Wardhaugh 2012

The moral rights of the author have been asserted

First Edition published in 2012

Impression: 1

All rights reserved. No part of this publication may be reproduced, stored in
a retrieval system, or transmitted, in any form or by any means, without the
prior permission in writing of Oxford University Press, or as expressly permitted
by law, by licence or under terms agreed with the appropriate reprographics
rights organization. Enquiries concerning reproduction outside the scope of the
above should be sent to the Rights Department, Oxford University Press, at the
address above

You must not circulate this work in any other form
and you must impose this same condition on any acquirer

British Library Cataloguing in Publication Data
Data available

Library of Congress Cataloging in Publication Data
Data available

ISBN 978–0–19–960542–2

Printed in Great Britain by
Clays Ltd, St Ives plc

For Jessica

CONTENTS

LIST OF ILLUSTRATIONS

Chapter 1

'Doctor Faustus' day'

MAKING IT FUN

> If thou do any knowledge gain thereby,
> Reader, thou art more wiser far than I.
>
> —Poor Robin's Almanac

Friar Tuck's Day: 23 June. Robin Goodfellow's: 31 October. Doctor Faustus': 14 April. As well as many more unlikely 'saints', the almanac for 1669 suggested that the reader commemorate the day—24 October—when the maid fell off the hen-roost, enjoy new cider in September, and celebrate forty-six years since the invention of beard-brushes. It advised that it would be dangerous to kiss a 'handsome wife' in front of her jealous husband.

This was *Poor Robin's Almanac,* written 'After a New Fashion' (evidently). If you had strolled into a bookshop in Restoration London, in the late autumn—say into Francis Kirkman's on Bishopsgate Street, conveniently close to the Royal Society's premises—you could have found a copy, and bought it for a few pennies. Kirkman had worked with Poor Robin, and although the almanac was small—a little over five inches by four, with just forty-eight pages—it was one of the bestsellers of the day, and must have cluttered his shop and many like it.

'If you like it', wrote the author, well and good; if not, 'do not prate of it', on pain of finding your name in next year's edition. At its peak in the 1670s, up to twenty thousand people liked it enough to buy a copy each year, ranging from Elias Ashmole, Fellow of the Royal Society and founder of the Ashmolean museum in Oxford, to forgotten working men and women for whom an almanac was—apart from the Bible—practically the only thing they would ever read. For them, an almanac might also be their only contact with the world of numbers and calculation, of astronomical language and astrological diagrams. This book tells their story: the story of ordinary people and the mathematics they knew, learned, used, read, and thought about, the books and teachers that brought it to them, and the things it did—and failed to do—for them. It is a story for which Poor Robin, who lived from the reign of Charles II to that of George IV, more than a hundred and fifty years later, is an ideal guide.

In this book we will see something of the very different ways mathematics could be used, of how mathematical calculations could go wrong and what the consequences could be, of where and how mathematics was learned, and of the beneficial effects that it was supposed to have on the mind and on the world. We'll also return to the theme of mathematics as a way to have fun, and to the tension which recurred throughout the eighteenth century between optimism and pessimism about what mathematics could or should do.

First, though, let's learn some more about Poor Robin and his world.

Poor Robin's world was the almanac. Small, convenient, and cheap, almanacs crowded the bookshops for a few weeks in the autumn and were bought, written in, used to destruction, and thrown away in huge numbers every year. In total, as many as four hundred thousand almanacs were printed each year in the England of Charles II and James II. One family in three bought one, and the almanac business was worth perhaps £2,000 per annum, a sum that would have employed (for example) rather more than a hundred able seamen.

Twenty or thirty or more different almanacs went on sale each year, and there seemed no limit to the information they could contain: not just a calendar of saints' days and new moons but a great deal more besides, catering to every imaginable shade of interest. If self-improvement was on the agenda, there were almanacs containing, say, a short course in solid geometry or a summary of classical mythology. For readers with specialized political or religious preferences there were almanacs giving historical and contemporary information of particular interest: lists of Royalist victories in the Civil War or discussions of the development of the English church.

Some almanacs specialized in medicine, and gave recipes for cures or for general-purpose tonics. Here's a remarkable prescription for a summer tonic given by Ferdinando Beridge in his almanac in 1654:

> Two gallons of morning-milk whey, one handfull of the herb Mercury, one of Mallowes, one of Violet Leaves, one of Cynck-foile... being all well boyled together (with some Licorish to relish it) and then clarified.

The hardy reader was supposed to drink 'halfe a Pinte at once', and promised that 'It will cleanse the body pretty titely, and save the purse'. We don't know whether anyone tried it.

Beridge worked in Leicester, and calculated the astronomical information in his almanac—sunrise, sunset, and the exact positions of the planets—so as to be correct for that location. While many English almanacs were calculated for London, and some came with the rather ambitious claim that they would 'indifferently serve' for the whole country, others, like Beridge's, were unabashedly regional. John Vaux, parson and astrologer, produced a Durham almanac for over forty years, and included in it not just astronomical but historical information tailored to that location. Accounts of the bishops of Durham and of Lindisfarne took the place of the more usual English worthies, and Vaux's table of historical dates was at least as concerned with 'The great fire in *Darlington*' and 'The building of the Free Grammar-School in Bishop *Auckland*' as with events of national or international importance.

Other almanac-writers aimed at audiences defined by sex, class, or profession. There were almanacs for women, for sailors, farmers, and brewers. Sometimes a single writer would be responsible, under different names, for as many as a dozen different almanacs each year. There were almanacs which specialized in legal information or in agricultural advice. And there were almanacs aimed at those interested in natural science. By the end of the century the 'Arch-conjuror' John Gadbury was indulging the vogue for such things with lengthy tables of weather observations (aimed at proving the widely-believed correlation of the weather with the phases of the moon) in his 'diary astronomical, astrological, meteorological', while Richard Saunders, 'Student in the Physical and Mathematical Sciences', included a short course in the mathematics of solid measurement in his almanac *Apollo Anglicanus,* as well as a 'discourse of the Motion of the Earth'. One, *The Royal Almanack,* adopted such a sober, learned tone as to earn itself a review—a favourable review—in the *Philosophical Transactions of the Royal Society.*

For readers with a sense of humour the almanac of choice was *Poor Robin.* In this crowded market it stood out: one of the most popular, and one of the most long-lived. When the fancy titles of the seventeenth century—*Apollo Anglicanus, Vox Stellarum,* and the rest—were forgotten, *Poor Robin* would roll on for 166 unbroken years. The first issue was printed in 1663, shortly after the Restoration of Charles II—the last came out in 1828, when the future Queen Victoria was eight. In that time it brought a ray of mathematical learning, and a good deal of fun, to hundreds of thousands of ordinary English people. *Poor Robin* rose, shone, and set during a time of change—perhaps the greatest there has been—in the way British people lived with and thought about mathematics, and in the role mathematics played in their lives. During a period which saw the

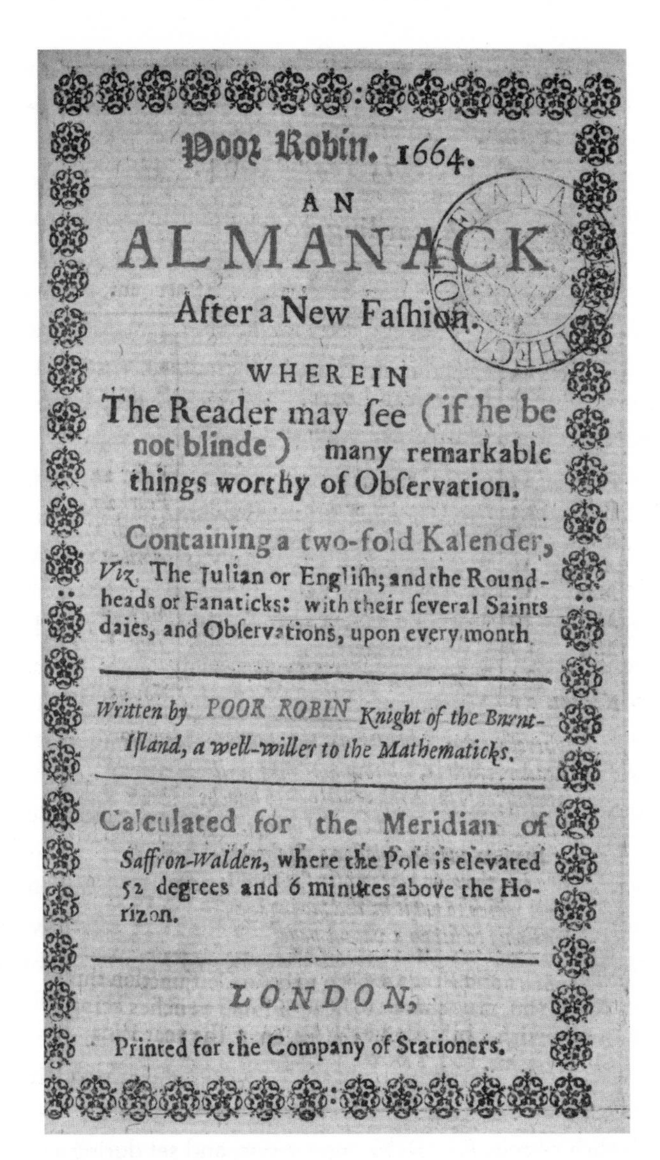

Poor Robin. 1664.

AN

ALMANACK

After a New Faſhion.

WHEREIN

The Reader may ſee (if he be not blinde) many remarkable things worthy of Obſervation.

Containing a two-fold Kalender, *Viz.* The Julian or Engliſh; and the Round-heads or Fanaticks: with their ſeveral Saints daies, and Obſervations, upon every month.

Written by POOR ROBIN *Knight of the Burnt-Iſland, a well-willer to the Mathematicks.*

Calculated for the Meridian of *Saffron-Walden,* where the Pole is elevated 52 degrees and 6 minutes above the Horizon.

LONDON,

Printed for the Company of Stationers.

FIG 1 The title page of Poor Robin's almanac, for 1664

The Bodleian Library, University of Oxford (Ashm. 586 (13))

beginning of persistent anxieties about the dehumanizing effect of a world of numbers and of 'facts, facts, facts', *Poor Robin* stood, on the whole, for the human face of mathematical learning—for the fact that mathematics could be fun and that numbers could be a tool in the hands of the powerless as well as the powerful.

How had things reached the point where almanacs were funny, where this particular exercise of mathematical skill could be the target of a (very) long-running parody? To answer this question we need to look just a little further back in time.

Britain in the seventeenth century was a period of frequent and drastic regime changes, when politics mattered and being seen to be on the right side was almost constantly important. Just a few years before *Poor Robin* made its appearance, Britain had been a kingless commonwealth. The licensing of the press meant that approved almanacs would necessarily be (mostly) those sympathetic to the commonwealth, but certain of the most prominent almanac-writers had aligned themselves with it very thoroughly, casting horoscopes for members of the Cromwellian regime and issuing predictions, interpretations, and downright prophecies openly sympathetic to its interests.

The most prominent of these was William Lilly, a man whose rise from poverty to fortune demonstrates the aspirations and success of the discipline of astrology at the time. His life, from 1602 to 1681, spanned a period of change in that discipline's reputation. He was a member of perhaps the last generation which could simply *do* astrology, without feeling obliged to get involved in an attack or defence of the subject. But he was also, by his sympathies, quite substantially responsible for making astrology seem inherently associated with puritanism and republicanism, thus, later, making it possible for *Poor Robin* to satirize it in the royalist-dominated 1660s.

So what did being an astrologer amount to? Popular astrology was, for many people, their only contact with mathematics and calculation, so if we wish to know about popular mathematics in this period, it's astrology we must look at. As it happens, we know a good deal about William Lilly's activities, although much of the evidence comes from his autobiography and thus reaches us through the lens of his own concerns and opinions.

He was the child of poor parents and went to a grammar school but not to university. He learned his trade from another astrologer, paying for a course of instruction which lasted seven or eight weeks. In that time he learned how to produce the characteristic diagram or 'figure' which showed in schematic form the positions of the planets at a particular moment, and he began to learn how to interpret such figures. He also, apparently, learned something about the magical conjuring of spirits, although he would later abandon that subject and claim to have burned his magic books.

After an argument with his first teacher, Lilly taught himself from books and eventually found another tutor. There was no such thing as a formal apprenticeship for an astrologer, but after some years of study supported by an inheritance he was ready to set out on his own as a professional. He was fortunate to find a patron to support him—William Pennington, a member of parliament—and he settled in London.

His private activities from about 1640 onwards consisted of further study based on a wide-ranging collection of sometimes obscure or rare books, and paid work on behalf of a diversity of individuals, from important politicians to those near the bottom of the social scale. He would, for instance, give astrological advice in medical cases, reporting what treatment was indicated by the stars, and perhaps by other divinations. On more than one occasion he was credited with saving a patient's life by his advice. In total he saw nearly two thousand private clients every year, ranging from servants to gentry, about half men and half women. Their questions ranged from detailed matters of political or military strategy, to the likely outcome of theological discussions, to mundane matters of love-life, family life, disease, and lost property. Prices ranged from half a crown to as high as £40. By 1662 he was earning as much as £500 a year: a very substantial income.

Lilly was also developing a more public side to his activities. The 1640s were a period when predictions of the future, particularly predictions of the future course of political events, were hungrily consumed by a concerned public. In a context in which many sectarian writers were predicting the imminent end of the world or cataclysmic events such as the fall of the Roman Church, cautiously worded prognostications about political matters could seem relatively measured and therefore relatively credible. Thus, one of Lilly's first published books was a fairly general reflection on the political and social implications of a forthcoming conjunction of Jupiter and Saturn.

More visibly in the long run, and perhaps more importantly for his developing image and that of English astrology, Lilly published almanacs:

> Merlinvs Anglicvs junjor The English Merlin revived, or, A mathematicall prediction upon the affairs of the English Common-Wealth, and of all or most kingdoms of Christendom this present yeer, 1644.

They appeared annually from 1644, and were from the beginning bestsellers, selling out within a few weeks even when their print runs reached tens of thousands. During the 1650s, Lilly's almanacs were translated into Dutch, German, Swedish, and Danish. His predictions had the fruitful ambiguity typical of the genre: a line or two alluding to an unspecified parliamentarian victory in June 1645, for example, enabled him to claim he had predicted both the occurrence and the outcome of the battle of Naseby.

As well as predictions, Lilly also issued pamphlets on specific subjects, including astrological interpretations of recent events and (he was becoming less cautious as a writer) sensational predictions of the imminent death of the king. These, too, were wildly popular, and Lilly now became valuable as a propagandist to the Parliamentarian cause. In time he was virtually the state astrologer—he was even one of the close committee to consult about the king's execution. But his strong words—and by this time they could be pretty strong—also earned him more than one brief spell in jail.

A final aspect of William Lilly's public activities was his production of the first important English-language textbook on astrology. *Christian Astrology*, published in 1647, was a deliberate attempt to make astrological knowledge available as widely as possible. In line with this aim, he also taught astrology—for a fee—to students in quite large numbers.

Through all of these activities William Lilly did a great deal to maintain and increase the visibility of his particular kind of 'mathematical learning' during the period of the Civil Wars and the Commonwealth, and to create a situation in which when people said 'mathematics' they most commonly meant astrology and the writing of almanacs. His fame rested in part on his ability to unify astrological divination both with religious prophecy and with natural philosophical knowledge, which he possessed in some quantity. One commentator notes that 'Lilly flourished at the last historical moment when such a thing was unselfconsciously possible'.

Like many other parliamentarians, Lilly survived the restoration of Charles II to his kingdoms in 1660. But although he retained his reputation—or at least his notoriety—he did not really prosper in the new political and social environment. His almanacs continued, but their sales declined, and he, and the subject of astrology, became less fashionable, even an object of ridicule. The association of astrology with puritan 'enthusiasm' and republican politics, for which Lilly had done so much, now inevitably worked against him. Indeed, astrologers of the generation after Lilly found themselves obliged to engage in a vigorous defence of their subject, while Lilly's name became a byword for its pretensions.

That said, he retained the support of some, for example of Elias Ashmole, who helped him gain a license to practice as a physician after the Restoration. Henry Coley, one of the new generation of astrologers, became his assistant and continued his almanac after his death. And at his death in 1681 he was, despite everything, a wealthy man, with more than sixty acres of land to his name. His trajectory ran from rags to riches on a scale that few professional mathematicians would ever match.

So it was thanks in part to Lilly that, when the Restoration came, almanacs, their writers, and by extension the whole of mathematics, came to be popularly seen as tarred with the brush of murderous Cromwellian politics and Puritan 'enthusiasm'. Hence Poor Robin's obsessive sallies against Puritan 'fanatics', and his willingness to lump all 'ass-trologers'—except himself—together as a pack of 'Thieves, Millers, Cheats, Quack-Doctors, Knaves, Impudent, Lying Astro-mongers... and hypocritical Nonconformists'.

Hence, also, the fact that the very word 'mathematician' came to mean not just 'astrologer' but 'liar'. A pamphlet published in 1660 illustrates this with a wonderful title addressing it to 'Eugenius Theodidactus, powder-monkey, roguy-crucian, pimp-master-general, universal mountebank, mathematician, lawyer, fortune-teller'. Learned individuals doing mathematical work called themselves, if they could, 'geometer' or 'professor of geometry', and although a 'mathematician' could be the innocent author of, say, a school primer, the word was for several decades just as likely to bear a satirical sense. The special phrase almanac writers used to describe themselves on their title pages was 'Well-willer to the mathematics', and it took until the 1690s to recover its serious reputation; for many years it could only be a sarcastic joke.

Into this situation burst Poor Robin, with a keen eye for what was ridiculous in a conventional almanac, but also a keen sense of what was useful and valued by its readers. His first almanac, for 1663, was printed illegally. Three thousand copies were sold before the law caught up with it, in the form of the Stationers' Company of London which, through its subsidiary the English Stock, held a monopoly on the production of almanacs. Having suppressed *Poor Robin* 1663 (not a single copy is left in the world today) the Company allowed it to reappear, with a proper license, the following year—and it never looked back. More than a century and a half later *Poor Robin*, now under the more affectionate title of *Old Poor Robin*, would be laid to rest when times had changed and its exuberance—and concomitant crudities—no longer seemed welcome.

Poor Robin's almanac looked *exactly* like a real almanac, down to the layout of the pages, their number, their size, and their shape. And it had largely the same contents: you could use it to find out moonshine like any other. Like any early modern almanac, it gave the sort of factual information which is still to be found in some pocket diaries. The dates of the full moons, times of sunrise and sunset, and other information about the movements of the planets, together with a calendar of the dates of the legal 'terms' and of major fairs and markets, and a long catalogue of saints' days and other ecclesiastical celebrations: Epiphany and Pentecost, the martyrdom of St Stephen and the birth of John the Baptist. There was a two-page spread for each month giving all this information, together with suggestions about when to plant your crops and harvest them, and some reflections, in dreadful verse, on the different months and seasons of the year:

> Now *sol* with warmer Raies smiles on the Earth,
> Giving the Fields and Woods a fruithful Birth,
> Of Cowslips, Peagles, Violets, Primroses,
> Wherewith young Country Lasses make them Poses.

Most almanacs also gave a 'prognostication' for the coming year, interpreting the movements of the heavens for their likely effect on the weather and human affairs. Year after year there were accounts of any coming eclipses of the sun and moon, notes on meteors and comets ('Blazing-stars'), the times of the equinoxes and solstices—and discussions of the earthly effects all of these might be expected to produce. The tone of this part varied from one almanac to another. Lilly continued to be spectacular and specific, of course, and plenty of people's political sympathies were still aligned with his. But many other writers were reluctant to follow him so far. Here's the rather cautious, not to say vague, discussion provided by Vincent Wing, the most popular almanac-writer of the Restoration period, concerning the solar eclipse of October 1668:

> which hapned in... *Scorpio*, the most viperus Sign in the *Zodiack*, as the learned observe, *Mars* claiming *chief domination*, and in the time of *greatest obscurity*, is unhappily posited upon the very *Cuspis* of the *Horoscope*, which according to the observations, both of the ancient and modern *Astrologians*, foreshews many *contentions, debates, quarrels, Duels*, much *underhand dealing, cozenage, disorders* and many *unruly actions* will ensue, and be manifested upon the *Stage* of the *World*.

Wing cited a host of ancient and modern authorities to magnify his prognostication: the science of astrology had a history all the way back to Babylon, and there were plentiful Greek and Roman sources on the subject. He concluded his almanac with a detailed discussion of the sizes and characteristics of the planets' orbits, taken from Copernicus and Kepler, whom he cited in the original Latin. He took pains, in other words, to appear very different from a sensational, perhaps rather vulgar, writer like Lilly. Where Lilly had taught astrology and carried out private horoscope readings, the Wings—father, son, and several later members of a dynasty publishing from the 1640s until the late eighteenth century—taught practical mathematics and worked as surveyors.

In addition to the political and religious differences which separated many astrologers of the Restoration from those of Lilly's generation, there were also two slightly different kinds of astrological prediction at work. Writes the historian Maureen Perkins:

> Judicial astrology was based on the premiss that a diagram of the heavens, a chart, could be drawn up and interpreted, giving the ability to interpret the significance of a moment in time and the events that the astrologer believed would follow that moment. Natural astrology applied information about planetary motion and lunar phase to physical phenomena such as the growth of crops or the administration of medicine.

Natural astrology (advising on when to plant and harvest your beans in terms of the year's cycle of seasons and moons) remained relatively respectable right into the nineteenth century, while judicial astrology (drawing up horoscopes) was the object of a good deal of scepticism, some of it deriving from William Lilly and the association of astrological divination with the Interregnum. As an example from the early eighteenth century, the 1728 *Chambers' Cyclopaedia* explained under its entry on 'Astrology' that judicial astrology was mere superstition, but natural astrology was soundly based on fact.

It is very difficult to generalize about how seriously either kind of prediction was taken. Critics were keen to suggest that only the very stupid read almanacs, but that was certainly not the case. At a time of constant new discoveries—or alleged discoveries—about how the world worked, and new theories about how correspondences between different parts of the world should properly be described mathematically or mechanically, perhaps stubborn scepticism had little to recommend it. A contemporary would not necessarily have understood why predicting the tides was different from predicting the weather, the plague, or the next war, or have perceived an essential difference between calculating the amount of light and heat to be received from the sun at a particular season and prognosticating the degree of light and heat to be expected in human affairs at that same season. Astrology offered a system of rules for understanding and predicting the world: one system among many, but a system with a pedigree, a large visibility in print, and a large following.

But almanac writers were keenly aware that some readers doubted what they had to say, while others would buy multiple almanacs each year in order to compare and contrast, and even write to complain to those who got things wrong. If some writers continued to indulge in highly, perhaps dangerously, specific prognostications (the more lurid and sensational the better—massacres, revolutions, deluges of blood, and so forth) many others, like Wing, confined themselves to generalities. Generalities which were, of course, easy for Poor Robin to mock.

In his almanac the joke was in the details, starting on the title page:

AN
ALMANACK
After a New Fashion.
WHEREIN
The Reader may see (if he be
not blind) many remarkable
things worthy of Observation...
Calculated for the Meridian of
Saffron-Walden, where the May-Pole is
elevated (with a Plumm cake on the top
of it) 5 yards $^{3}/_{4}$ above the Market-Cross.

That quasi-mathematical precision about the maypole was a trademark. Normal almanacs would always declare where they were calculated for, giving the 'elevation of the pole'—the latitude, or rather the distance in latitude from the North Pole—at that location. Poor Robin and other satirists placed themselves all over the globe and beyond, putting the pole 'many miles above sense and apprehension', or even 'an hundred and fifty degrees above all reason'.

Poor Robin's table of notable dates was conventional enough: the Gunpowder Plot, the coronations of Charles I and Charles II, the Plague and the Great Fire. But on the facing page the dates continued in a slightly different mood: 'Women first invented kissing' (1367 BC); 'The invention of lying' (3934 BC); or, charmingly,

'Curran[t]s were first put in Cakes' (AD 1164). It was the same with the list of saints' days in the calendar. On one page, St Julian, St Christopher, St Polycarp and the Conversion of St John. On the facing page, Don Quixote, Robin Hood, Mother Shipton, and 'Jane fell off the Hen-roost'. Poor Robin seemed inexhaustible when it came to this kind of thing, tirelessly turning out unexpected conjunctions from fiction, folklore, and the real world (Joan of Arc, Doctor Faustus, Patient Griselda) together with the tiny joys and setbacks of Restoration life. Twenty-eighth of June 1670: 'Curds and cream you never saw the like on't'.

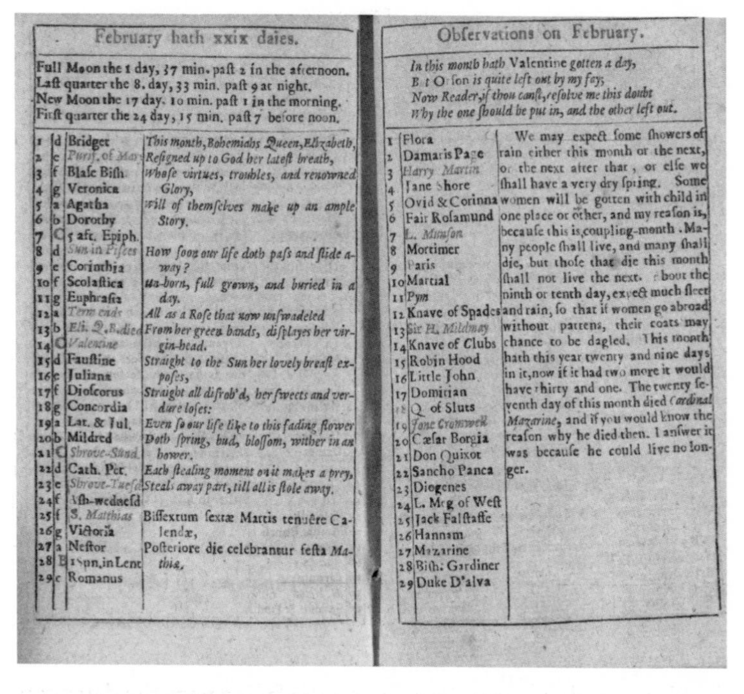

February hath xxix daies.

Full Moon the 1 day, 37 min. paſt 2 in the afternoon.
Laſt quarter the 8. day, 33 min. paſt 9 at night.
New Moon the 17 day. 10 min. paſt 1 in the morning.
Firſt quarter the 24 day, 15 min. paſt 7 before noon.

1	d	Bridget	*This month, Bohemiahs Queen, Elizabeth,*
2	c	Purif. of Mary	*Reſigned up to God her lateſt breath,*
3	f	Blaſe Biſh.	*Whoſe virtues, troubles, and renowned*
4	g	Veronica	*Glory,*
5	a	Agatha	*Will of themſelves make up an ample*
6	b	Dorothy	*Story.*
7	C	5 aft. Epiph.	
8	d	Sun in Piſces	*How ſoon our life doth paſs and ſlide a-*
9	e	Corinthia	*way?*
10	f	Scolaſtica	*un-born, full grown, and buried in a*
11	g	Euphraſia	*day.*
12	a	Term ends	*All as a Roſe that now unſwadeled*
13	b	Eli. Q. B. died	*From her green bands, diſplayes her vir-*
14	C	Valentine	*gin-head.*
15	d	Fauſtine	*Straight to the Sun her lovely breaſt ex-*
16	e	Juliana	*poſes,*
17	f	Dioſcorus	*Straight all diſrob'd, her ſweets and ver-*
18	g	Concordia	*dure loſes:*
19	a	Lat. & Jul.	*Even ſo our life like to this fading flower*
20	b	Mildred	*Doth ſpring, bud, bloſſom, wither in an*
21	C	Shrove-Sund.	*hower.*
22	d	Cath. Pet.	*Each ſtealing moment on it makes a prey,*
23	e	Shrove-Tueſd.	*Steals away part, till all is ſtole away.*
24	f	Aſh-wedneſd	
25	f	S. Matthias	Biſſextum ſextæ Martis tenuêre Ca-
26	g	Victoria	lendæ,
27	a	Neſtor	Poſteriore die celebrantur feſta *Ma-*
28	B	1 Sun. in Lent	*thiæ.*
29	c	Romanus	

Obſervations on February.

In this month hath Valentine *gotten a day,*
But Orſon *is quite left out by my fay;*
Now Reader, if thou canſt, reſolve me this doubt
Why the one ſhould be put in, and the other left out.

1	Flora
2	Damaris Page
3	Harry Martin
4	Jane Shore
5	Ovid & Corinna
6	Fair Roſamund
7	L. Munſon
8	Mortimer
9	Paris
10	Martial
11	Pym
12	Knave of Spades
13	Sir H. Mildmay
14	Knave of Clubs
15	Robin Hood
16	Little John
17	Domitian
18	Q. of Sluts
19	Jone Cromwell
20	Cæſar Borgia
21	Don Quixot
22	Sancho Panca
23	Diogenes
24	L. Meg of Weſt
25	Jack Falſtaffe
26	Hannam
27	Mazarine
28	Biſh: Gardiner
29	Duke D'alva

We may expect ſome ſhowers of rain either this month or the next, or the next after that, or elſe we ſhall have a very dry ſpring. Some women will be gotten with child in one place or other, and my reaſon is, becauſe this is coupling-month. Many people ſhall live, and many ſhall die, but thoſe that die this month ſhall not live the next. About the ninth or tenth day, expect much ſleet and rain, ſo that if women go abroad without pattens, their coats may chance to be dagled. This month hath this year twenty and nine days in it, now if it had two more it would have thirty and one. The twenty ſeventh day of this month died *Cardinal Mazarine*, and if you would know the reaſon why he died then. I anſwer it was becauſe he could live no longer.

FIG 2 A page spread from Poor Robin's almanac for 1664, showing the two versions of the calendar, serious and comic

The Bodleian Library, University of Oxford (Ashm. 586 (13))

There is something very appealing about the world Poor Robin's almanac evokes, with its perpetual merry round of Don Quixote's birthday and curds and cream: 'Peter broke his shins at foot bal', Robin Goodfellow's day and 'My man *Tom* was drunk and neglected going to Plow'. In 1669 he remarked that he had probably had as much pleasure writing it as the reader in reading it, and that fact is obvious from the jovial tone and (usually) gentle humour.

But it did come at a price. If you were not wholly loyal to the Stuart monarchy and the established church, you were not a participant in this charming world but one of Robin's satirical victims, a 'fanatic', a 'maggot-head', one of those 'who brought our *Royal Shepherd* to the Block', and whose list of commemorations included St Oliver (Cromwell) and his blood-soaked henchmen alongside Mephistopheles, Bluebeard, and Pontius Pilate.

Poor Robin spoke to a generation that had lived through the upheavals of the Civil Wars and Interregnum, and he was in no doubt of his opinion of the events of those times. During the troubled period of the 1680s he would remain staunchly loyal to the established order, be it Stuart or Orange. In the crucial year of change the long lead-time involved in producing the almanac caused some embarrassment for him, and for other prognosticators: the text of the almanacs for 1688 had to be ready by the spring of the previous year, and Poor Robin thus appeared in print with good wishes to James II somewhat after that unhappy monarch had fled the country and the throne.

Such difficulties aside—and there would be no upheaval directly affecting Robin in later life comparable to the revolution of 1688–9—the vicissitudes of politics had little real effect on him. More interested in maintaining his own popularity than in following a fixed path, he leaned towards a gentle Toryism, sympathetic to the interests of Crown and country as against city and parliament, but stopping well short of offensive enthusiasm and with never a whisper of support for the exiled Stuarts, still less the papists. Poor Robin was, throughout his long life, a wise fool, a jester—his defining characteristic was that he knew better than his betters, and wasn't afraid to say so. But his message was one of mockery within acceptance of one's lot, not of revolution. If the level of explicit political commentary varied from year to year, depending in part on the political temperature (it was generally rising during the 1660s and 1670s), it was always safe, and it could be distinctly lumpish:

> This month dy'd Cromwell in good sober sadness,
> Whose death fill'd al England with much mirth and gladness ...
> What became of his soul (when his body did stink)
> There is no man can tell, but I know what I think.

Like many an early modern almanac, *Poor Robin* also gave some mathematical information, in the form of ready reckoners. A table of interest showed the sums that would have to be repaid if a given amount of money was lent at a rate of 6 per cent for one month, two months, or more. 'Money it is a strange property / Though dead, it doth encrease and multiply.'

For it wasn't all bad news for mathematics and its reputation during this period. At the Restoration, Charles II's triumphal procession through the streets of London on the way to his coronation featured, among many other things, the personifications of arithmetic, geometry, astronomy, and navigation, arranged in niches on a triumphal arch. They took the form of women variously attired ('*à l'antique*') and bearing such attributes as a book full of numbers, a compass, a triangle, stars, and an anchor, the designer perhaps having been at a loss for an easily comprehensible symbol of the navigational art. The mathematical arts were vital supports of public and private life, as we will see again and again in this book, and Lilly and his colleagues had done nothing to change that.

The great education reformer Samuel Hartlib, for example, had proposed in 1653 that

> every one should be able to reckon things by numbers, and measure them by their weight & dimensions of breadth and length, and height and depth iustly [justly], that they may not be cozened.

So Poor Robin—as well as making fun—saw himself as performing a vital altruistic service in his almanac and his other writings. In the straightforward morality of Robin's jolly king-and-country world, the poor were constantly beaten down by the rich yet constantly finding ways to get their own back, and in this context Poor Robin's unique brand of common sense and wry honesty was a tool which could help them avoid being deceived:

> We shew no Jugling-Tricks, nor idle Fancies,
> To fill the Ignorant People with Romances;
> All the Predictions which we do fore-tell,
> Are as unquestion'd as an Oracle.

By buying into Poor Robin's world you learned a wariness of quackery which might stand you in good stead in the hard world of 'eclipses of money' and scolding wives, embarrassing accidents and humiliating diseases. One of many non-almanac publications produced by the character of Poor Robin was a *Character Of an Unconscionable Pawn-Broker*. Putting Robin's mathematical expertise at the service of the indigent and gullible, it pointed out that a pawnbroker's effective rate of interest could amount to 'thirty three pounds, six shillings, and eight pence in the hundred', nearly five times the rate of interest allowed by law on a loan of money. Mathematics was a dangerous weapon, and Robin was concerned to put his readers on the right side of it.

Who was Poor Robin? As every title page proclaimed, he came from Saffron Walden in Essex. According to his own works he was born, it seems, around 1620. His strong views about the civil wars and commonwealth (and about everything else) had been acquired through first-hand experience. Scattered references to a folklore character called 'Poor Robin' can be found in ballads from the 1640s and 50s, but it was not until the 1660s that he became highly visible.

After many years' study of astrology he launched himself on an unsuspecting public in 1663, when he published both his first almanac and *The Path-Way to Knowledge*, a longer book about astrology which, in the best Poor Robin style, borrowed a title previously used by arithmetic textbooks. By the end of the century he was the author of nearly twenty more books and pamphlets, as well as thirty-eight annual editions of his almanac. Some made no secret of their comic intentions, like *Poor Robin's Prophecy* (1701) 'carefully calculated, to make both sexes shake their sides till they break their twatling-strings' (the dictionaries know nothing of 'twatling-strings', but it was evidently intended to sound vaguely rude). Others hid their merriment behind veils which have deceived casual observers to the present day. In modern library catalogues we can find, for instance, Poor Robin's uproarious *Prophesies and divertisements* (1677) catalogued under 'prophecies, occultism and fortune-telling'.

As well as his books of 'prophecies', Poor Robin produced 'characters' of Frenchmen, Dutchmen, and 'honest drunken curts', topical political pamphlets, bawdy stories and volumes of 'jests'. He even wrote a weekly newspaper—*Poor Robin's Intelligence*—which ran for about 120 issues between 1676 and 1680. Rumour claimed that a 'whole Bank' of 'Slaves' was involved in the production of the *Intelligence*, but Robin denied it.

Robin seems to have held his headquarters in the Queen's Head tavern on Snow Hill in the heart of London, and during the period of his penny newspaper he invited members of the public to seek him there if they had stories to pass on. Once or twice he hinted that he was also taking money to show people the stars or teach them astrology—he wouldn't have been unusual in doing that, but it's hard to imagine who would have paid for his lessons, or what he would have taught them.

The zenith for Poor Robin-related publications came in the 1670s, when he produced ballads, biographies, and accounts of his journeys, and references to him in fiction, poems, and plays are almost innumerable—for a while he became a byword for mock astrology. According to his own estimation, by the mid-1670s the fear of ridicule in his works had become so great that he '[kept] the peace no less than the *Constable*', exposing 'Vice and *Ill Nature* to deserv'd contempt' and 'if possible Laugh[ing] *Foppery* out of Countenance and Practice'. Others, naturally, were less convinced, and there were public attacks from those he had offended. *A scourge for Poor Robin* appeared in 1678 and *Poor Robin turn'd Robin the devil* in 1680, while his unfavourable remarks about the nonconformist minister Thomas Danson provoked a pamphlet from him and a reply from Robin during 1676–7: Robin's 'Natural Antipathy and *contempt*' against Danson's very violent indignation. They were two '*Fencers* in Divinity', as Robin had it, 'who like *Tavern Bullies* catch up whatever stands next, and convert it into a Weapon... for carrying on a *Brangle*'.

So great was Robin's fame that he also became a victim of fraud. The temptation to attach the Poor Robin name to works not approved by the original author was overwhelming, and while some of these were benign imitations, others tried to make Poor Robin serve ends uncongenial to him. Uncertainty continues to this day. Not all of Robin's works have survived, and rumours that he—or an imitator—was an author of pornography in the 1670s and 1680s seem difficult to confirm or deny.

But for his supporters Poor Robin could do little wrong—his every work luminous with mirth and pleasure (not to mention social and political commentary). A false report of his death in 1679 brought forth this epitaph from an anonymous admirer:

Here lies *Poor Robin,* most enriched one
With *Nature's* Dowre, *Graces* large Portion.
Nature brought *Reason, Prudence, Eloquence,*

And *Magnanimity*, *Munificence*,
Courage and *Constancy*, and Matchless *Wit*.

His very best work was in some of the 'prognostications' in his early almanacs, where the self-serving safety of some astrologers' predictions prompted him to play his wry games with fact and fiction to the full. When Wing solemnly advised the readers of his almanac that 'about the end of *January*, and middle of *February* 1671' they should 'expect much *Snow* or *Rain*, and bad Weather', Robin retorted with a prediction of 'Sharp weather and hard Frosts...in *Green land*'. He delighted in similar fail-safe prophecies:

> If on the second of *February*, thou go either to Fair or Market with store of money in thy pocket, and there have thy purse picked of it all, then that is an unfortunate day.

> It is to be thought some people will die this month in one place or other: and if there be no Wars in the World, we are like to have a very peaceable one.

> Two in a bed will be warmer than one.

Poor Robin of Saffron Walden, of course, didn't really exist. Over his long life several different writers were responsible for him, but for the most part they are not even names to us. And nobody really knows who was behind some of Poor Robin's works in his heyday of the 1670s and 80s. But his original creator was himself a well-known man in his day. His name was William Winstanley, and he lived from about 1628 to 1698. A freeman of Saffron Walden and resident of the nearby village of Quendon, he began his publishing career in 1655 with a volume of poems. Over the next forty-odd years his output would range over biography, astrology, history, and more. By the middle of his career he would claim to have written 'above seven-score books'. He didn't invent the parody almanac (there had been spoof astrologers since ancient times, and Rabelais had written a marvellous *Pantagrueline prognostication* in the 1530s, possibly the earliest printed example of the genre), but by creating *Poor Robin* he gave it new life whose success he could hardly have imagined. By Winstanley's death in 1698, *Poor Robin* was established as part of the publishing landscape, and would remain popular when his creator was long forgotten.

We don't know much about Winstanley's background or his family (although there could be a connection with the slightly younger Henry Winstanley, creator of the first Eddystone Lighthouse, born in Littlebury near Saffron Walden). We do know that he was a true-blue Englishman of his generation—he'd lived through the Civil War and the Interregnum and the only thing he hated more than a Puritan fanatic was a Papist. Throughout his output, he pulled no punches in his promotion of Royalist politics and Anglican conformity. Of the poet John Milton—'a notorious Traytor'—he wrote that 'his Memory will always stink'. As 'Philoprotest' he produced a *Protestant Almanac* from 1668 until his death, and anonymously an *Episcopal Almanac* (against Protestant Dissenters) and a *Yea and nay Almanac* (against Quakers). None attained anything like the success of *Poor Robin*, with which their tone of credulous illiberality, and their dullness, make an unhappy contrast—the last two continued for only a few issues each.

While much of the early Poor Robin material was Winstanley's work, the weekly newspaper—*Poor Robin's Weekly Intelligence*, and various revivals—seems to have been produced by another man. Henry Care, a self-taught expert in religious affairs, law, and history, put his considerable knowledge of languages to use in a flood of translations, pamphlets, and other writings throughout the 1670s and 1680s. His modern biographer finds him 'guilty of lying, plagiarizing, and libelling' and notes that his writings 'helped to embed anti-Catholic prejudice in the national consciousness'.

Although Care's Poor Robin work was rather more politically innocuous than the bulk of his output, he cannot have been a congenial figure to Winstanley, whose stolid politics stand in conspicuous opposition to Care's mercurial (some would say opportunistic) support for both Protestant Dissenters and the Catholicizing policies of James II. We don't know anything about relations between the two men, which could have been minimal.

Care was less witty, too. Nearly all of what the *Intelligence* contains is bawdy: its world was one in which there was really only one joke, and that joke was *horns*. The 'news' generally took the form of descriptions of alleged crimes or political and military reports from imaginary locations, and consisted largely of stories about cuckoldry, interspersed with a rather heavy topsy-turvydom in which women got the better of men, servants of their masters, and the poor of the wealthy.

Mathematics wasn't especially visible in the newspaper, but it was there. Poor Robin would have been known to readers as the writer of the funniest almanac of the age, and they might have read it with that in mind. When mathematics did appear in this world it was one form of expertise among many and, like others, its main function was to oppress the poor and powerless, as for instance in the person of greedy landlords on rent day—the start of the legal year—in 1676:

> Landlords and Tenants talk of nothing now but disbursements, Leases, and Releases, Payments, Quit-Rents, Discharges, &c. The Lawyer and his Clyent of Mortgages, Principal and Interest: but, in case there arise any dispute, the Clyent... spends as much in defending the controversie as would have ended it. According to the method of some Cooks; who make you pay just so much for dressing as your victuals cost.

Like medics, public officials, and the heads of families, people with numerate expertise came in for heavy censure for deceiving those too ill-educated or credulous to defend themselves. Weighing and measuring could be metaphors for other kinds of dishonesty: 'taking measure of her with more than ordinary familiarity', and 'she was only practising the Mathematicks'.

Practical mathematicians were fairly common in this world, often in the form of members of Robin's own profession—quack astrologers—who in one way or another got their comeuppance. Elite mathematicians came in for censure too. Care, on one occasion, had Robin pour scorn on a new mathematical discovery, an alleged squaring of the circle: 'a Non existent Entity, or a being that is not'.

Numerical and geometrical activities were again and again made out to be inherently dishonest, as metaphors for lying and adultery—the preserve of avaricious landlords, unscrupulous merchants, and mendacious astrologers. Throughout the newspaper they were virtually a synonym for deception, for the presentation of nonsense as sense: in Robin's words, 'puzzling the intellects and beguiling the Fancies of the easie and credulous'.

Care's was a more direct attack on astrology, and a more broadly-directed, less subtle—less intelligent—abuse of mathematics and mathematicians than Winstanley's, even if it made some of the same points and employed some of the same language. Behind the Poor Robin image there evidently lay powerful personal differences, but the writing never failed to continue even in the most politically troublesome days of the 1680s.

The lives of these hack writers cannot have been enviable. Simply writing the quantity of text Winstanley or Care produced would have been a gruelling, even debilitating labour in the days of dip pens and candlelight, and we hear of writers driven to blindness, insanity or death by the meagre conditions of life which their efforts enabled them to afford. Furthermore, these were the days before effective copyright protection, when what you could publish was, to some extent, what you dared and when the retribution of disgruntled readers or writers could be violent and drastic (the bookseller Abel Roper came to blows with the writer Tom Brown in 1700, and won; later the poet Alexander Pope was rumoured to have spiked the publisher Edmund Curll's drink with an emetic). *Poor Robin* was produced in difficult times, and we can forgive his writers their differences.

Plagiarism was as common in the almanac business as elsewhere, despite the efforts of the Stationers' Company to enforce some sort of regulation, and several almanacs regularly carried a warning to accept no deleterious imitations. Things reached a crisis in 1698 when William Winstanley died, probably in his sixties. He had completed the almanac for 1699, but what was to happen the following year? Poor Robin could not possibly be allowed to die with him, but who was going to take up the baton?

The—perhaps predictable—result was that in 1700 *two* versions of Poor Robin's almanac were printed. One had the authorization of the Stationers' Company of London, the other lacked it and was printed—or claimed to be printed—in Dublin (it would be decades before any form of copyright protection was extended to Ireland, and Dublin was a centre of book piracy throughout much of this period). Both were anonymous, but the Dublin one seems to have been the work of 'that Eminent *Astrologer and Physician*, Mr. *John Whalley*'. His imitation was completely convincing—he caught Poor Robin's tone perfectly, and if we didn't have the London version to compare it with the fraud would be very hard to spot.

This situation could have been serious for Robin's long-term health, since neither of the authors—whoever they were—could make any obviously plausible claim to be the 'real' Poor Robin, or rather the legitimate inheritor of the Poor Robin tradition. He was hardly the sort of thing Winstanley could have left to someone in his will.

Yet the situation resolved itself, perhaps with the intervention of the Stationers' Company. Whatever success his version of *Poor Robin* met with, Whalley didn't produce another one the next year, and the official author for 1701 didn't deign to mention the incident directly, only obliquely noting that 'We hunt not after popular Applause / Or foraign Praise'. There would be further imitations, as we'll see, but for now there was just one official version of *Poor Robin* on sale. We don't, unfortunately, know who was writing it.

Poor Robin wasn't the only person making fun of mathematics and mathematicians in Restoration Britain. He had colleagues both inside and outside the almanac business.

Tom of Bedlam was descended from the 'Poor Tom' of *King Lear* (Edgar in disguise, then, but by this stage he was very thoroughly disguised). The character was a popular one, it seems, and several early

FIG 3 Frontispiece from *News from Bedlam*, 1674
The Governing Body of Christ Church, Oxford (4.B.208)

editions of the play gave him nearly as much space on their title pages as Lear himself. In (some) early performances he sang a 'mad song', to which wags were quick to fit other words, even wholly inappropriate ones, an early example being the anonymous 'Cunning Northerne Begger' of 1634. Thus Poor Tom took on a life of his own.

That life reached a point—and a distance from Shakespeare's immortal tragedy—where in 1674 it made sense to someone to make Poor Tom the author of *News From Bedlam*, a satirical almanac 'Calculated chiefly for the meridian of Great Bedlam . . . where the pole is elivated many miles above sense or apprehension.' This production was almost worthy of Poor Robin himself—he even got a mention at one point—although Tom was more concerned to attack astrology than to make fun. Robin was not impressed—he claimed that Tom's almanac was mostly copied from his own, and crowed over its failure to continue to a second edition.

On the title page Tom was sublimely described as the 'Knight of the Frantic Horn', 'student in mathematical gimcracks, whimsies, anticks, and others rare chymera's'. 'Gimcrack' was also the name given to the scientist Robert Hooke in a satirical play about him a couple of years later. The wonderful frontispiece shows Tom with his mathematical instruments. (If you're wondering, he's holding a quarterstaff, for finding the elevation above the horizon of an object in the sky. Tom seems to be pointing it at the moon and its bibulous inhabitant, but unfortunately the attached quadrant has no markings on it and would be useless for reading off the elevation. He also has what looks like a globe of the heavens, with a few stars marked on it. The device in the bottom right of the picture—two horns and a triangle—was meant for Tom's coat of arms.)

'Merry Andrew' had an equally colourful back story. He burst into notoriety in 1672 in a 'jovial discourse' with Poor Robin ('The colledge of fools display'd and their capps tost at tennis') and several other works, and seems to have been involved with a popular puppet theatre. He made more than fifty further appearances in print by the end of the century, cropping up in works by John Dryden, Aphra Behn, Daniel Defoe, and Andrew Marvell, to name just a few. In 1699, as 'a well-wisher to the mathematicks', he became the author of a mildly whimsical Edinburgh almanac in 1699, borrowing a few titbits from *Poor Robin*, and possibly attempting to steal the market from the ailing Winstanley. The almanac didn't last, but the character did, and 'Merry Andrew' is still in the dictionaries, if not in very common use any more, as a name for 'a fool, an idiot; a joker'.

One of the best of Poor Robin's satirical colleagues was Sidrophel, the astrologer in Samuel Butler's hilarious spoof epic *Hudibras*, published in the same year, 1663, as Poor Robin's first almanac. Hudibras himself was a fanatical puritan 'knight' in the mould of Don Quixote, with a 'Skull / that's empty when the Moon is full'. He had some mathematical competence of his own:

> For he, by geometric scale,
> Could take the size of pots of ale;
> Resolve, by sines and tangents straight,
> If bread or butter wanted weight;
> And wisely tell what hour o'th'day
> the clock does strike, by Algebra.

But when the going got tough he made the unsurprising decision to consult an astrologer, and Sidrophel stepped into the frame. He was, predictably, a satire on William Lilly:

> has not he point-blank foretold
> Whats'e'er the close committee would?
> Made Mars and Saturn for the cause,
> The Moon for fundamental laws,
> The Ram, the Bull, the Goat, declare
> Against the book of common prayer?

Yes, he was a puritan fanatic. And he pulled the wool over Hudibras' eyes at first. 'It is no part / Of prudence to cry down an art, / And what it may perform, deny, / Because you understand not why'.

Sidrophel and his assistant Whachum blew their cover, though, by attacking Hudibras, and things ended badly. All the same, Sidrophel was a learned man, in his way:

> He had been long t'wards mathematics,
> Optics, philosophy, and statics,
> Magic, horoscopy, astrology,
> And was old dog at physiology.

And

> He made an instrument to know
> If the moon shine at full or no;
> That would, as soon as e'er she shone, straight
> Whether 'twere day or night demonstrate;
> Tell what her d'ameter to an inch is,
> And prove that she's not made of green cheese.

We could multiply examples. Margaret Cavendish, Duchess of Newcastle and a published author on natural philosophy, for example, wrote a utopian fantasy in which she depicted geometers as dealing with imaginary, incomprehensible things and astronomers as a danger to the state due to the violence of their arguments:

> The Spider-men [her mathematicians] came first, and presented her Majesty with a table full of Mathematical points, lines and figures of all sorts of squares, circles, triangles, and the like, which the Empress, notwithstanding that she had a very ready wit, and quick apprehension, could not understand, but the more she endeavoured to learn, the more was she confounded. Whether they did ever square the circle, I cannot exactly tell, nor whether they could make imaginary points and lines, but this I dare say: That their points and lines were so slender, small and thin, that they seemed next to Imaginary.

But it is time to move on. Almanacs weren't the only way to hear about mathematics in early modern Britain, and they weren't the only place in which optimism, pessimism, eccentricity, and satire on the subject were played out. We'll see much more of all those things in the remainder of this book, as we journey through the worlds of education, hard work, self-improvement, recreation, and much more, and Poor Robin will continue to provide his wry take on the story. In the next chapter we'll continue our story into the early eighteenth century, and look at some of the ways in which mathematics could affect people's lives in early modern England.

Chapter 2

'The dismal and long expected morning'

GETTING IT WRONG

The streets of London were crowded. Fellows of the Royal Society had set up their telescopes at Gresham College on Bishopsgate Street, as had the astronomers at the Royal Observatory at Greenwich. Old women wore special spectacles, young men prepared themselves buckets of water so as to watch the remarkable phenomenon in the dubious safety of a reflection, and the Eton schoolboys made pinhole cameras out of their top hats.

This was the morning of Wednesday, 13 September 1699. A solar eclipse was predicted, and since as much as eleven-twelfths of the sun was expected to be obscured, London made very elaborate preparations. Lamplighters were on hand to provide illumination, should it become necessary, and 'link-boys' (torch-carriers) were ready 'to light Gentlemen to the Tavern at mid-day'. Balconies and even church steeples were crowded, and the London Monument 'was bespoke a Fortnight before, and places there went at Half a Crown a Man.'

Outside the capital there was even greater excitement. The almanac makers had done their predictable worst, and preparations for the eclipse approached the downright hysterical, with the price of corn reportedly rising four pence per bushel in some provincial markets in anticipation of shortages. Pious women were reported to have sat up all night, praying.

'At last the dismal and long expected morning came'. But although 'the whole Town waited from Seven in the Morning, till one at Noon' there was nothing—or apparently nothing—to be seen. All preparations were void, and eventually the citizens went to their dinner, 'as angry with the Sun and Moon for bilking them, as they were concern'd for them before'.

These reports are from contemporary newspapers, and although they suffer from a certain exaggeration ('An old Gentleman was so concern'd for the matter, that he read the Service for the Churching of Women in the Common Prayer Book, and applied it to the Sun') they seem to be true in essence. Anticipation of the eclipse overshadowed England for a week or so in autumn 1699, reaching its darkest for a few hours on the morning of the eclipse itself. The public mood was blackened still further—apparently on purpose—by almanac makers, newspaper writers, and the authors of a small flood of cheap astronomical publications in the weeks before the eclipse (Figure 4 shows the wonderful illustration from one of those pamphlets). When the great day came, though, it all went wrong. The weather was fine, but the predictions were out to such a degree that an eclipse which was supposed to be virtually total produced almost no visible effects at all. Modern calculations suggest that from London about 85 per cent of the sun was covered by the moon, which would have shown up with a pinhole camera but would not have made the world seem noticeably dark.

The incident fits neatly into a story of the declining credibility of almanacs in early modern England and the rising currency of their satirical cousins. 'The Country People tore their Almancks in a rage, and will not believe those Gentlemen for the world, tho' they Prophesy Snow in *January* and hot Weather in the Dog-days'. You'll recall that some almanacs really did confine themselves to

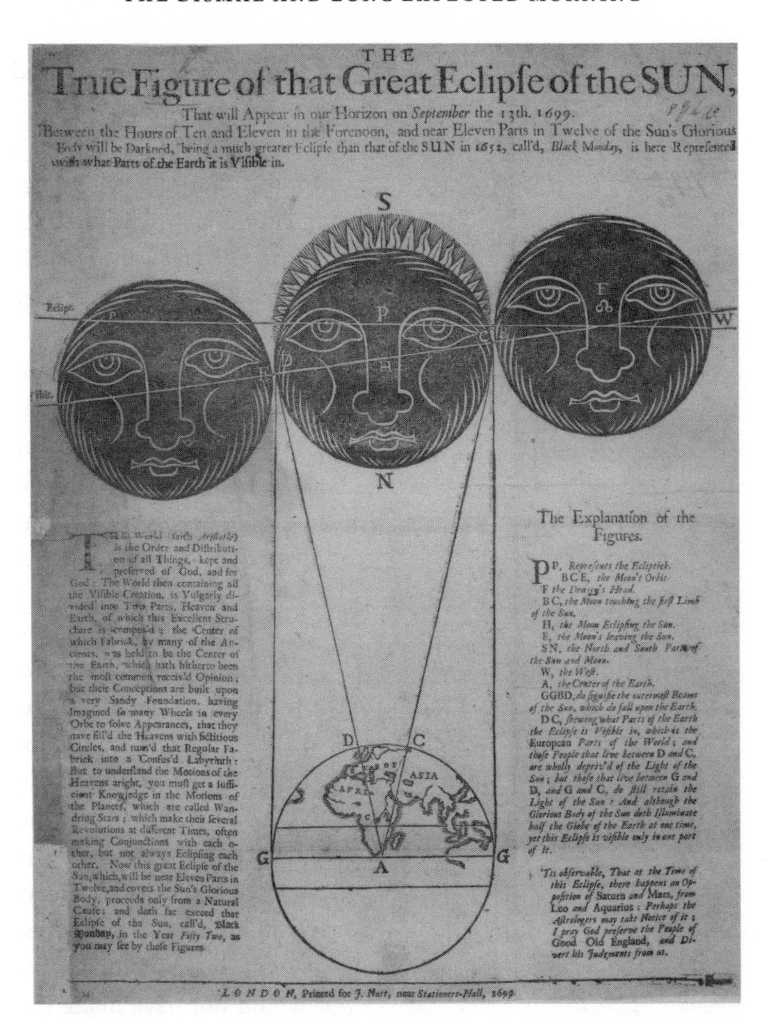

THE
True Figure of that Great Eclipſe of the SUN,

That will Appear in our Horizon on *September* the 13th. 1699.
Between the Hours of Ten and Eleven in the Forenoon, and near Eleven Parts in Twelve of the Sun's Glorious Body will be Darkned, being a much greater Eclipſe than that of the SUN in 1652, call'd, *Black Monday*, is here Repreſented with what Parts of the Earth it is Viſible in.

THE World (ſaith *Ariſtotle*) is the Order and Diſtribution of all Things, kept and preſerved of God, and for God: The World then containing all the Viſible Creation, is Vulgarly divided into Two Parts, Heaven and Earth, of which this Excellent Structure is compos'd; the Center of which Fabrick, by many of the Ancients, was held to be the Center of the Earth, which hath hitherto been the moſt common receiv'd Opinion; but their Conceptions are built upon a very Sandy Foundation, having Imagined ſo many Wheels in every Orbe to ſolve Appearances, that they have fill'd the Heavens with fictitious Circles, and turn'd that Regular Fabrick into a Confus'd *Labyrinth*: But to underſtand the Motions of the Heavens aright, you muſt get a ſufficient Knowledge in the Motions of the Planets, which are called Wandring Stars; which make their ſeveral Revolutions at different Times, often making Conjunctions with each other, but not always Eclipſing each other. Now this great Eclipſe of the Sun, which, will be near Eleven Parts in Twelve, and covers the Sun's Glorious Body, proceeds only from a Natural Cauſe; and doth far exceed that Eclipſe of the Sun, call'd, **Black Monday**, in the Year *Fifty Two*, as you may ſee by theſe Figures.

The Explanation of the Figures.

PP, *Repreſents the Ecliptick.*
BCE, *the Moon's Orbit.*
F *the Dragon's Head.*
BC, *the Moon touching the firſt Limb of the Sun.*
H, *the Moon Eclipſing the Sun.*
E, *the Moon's leaving the Sun.*
SN, *the North and South Parts of the Sun and Moon.*
W, *the Weſt.*
A, *the Center of the Earth.*
GGBD, *do ſignifie the outermoſt Beams of the Sun, which do fall upon the Earth.*
DC, *ſhewing what Parts of the Earth the Eclipſe is Viſible in, which is the* European *Parts of the World; and thoſe People that live between* D *and* C, *are wholly depriv'd of the Light of the Sun; but thoſe that live between* G *and* D, *and* G *and* C, *do ſtill retain the Light of the Sun: And although the Glorious Body of the Sun doth Illuminate half the Globe of the Earth at one time, yet this Eclipſe is viſible only in one part of it.*

'Tis obſervable, That at the Time of this Eclipſe, there happens an Oppoſition of Saturn *and* Mars, *from* Leo *and* Aquarius: *Perhaps the Aſtrologers may take Notice of it; I pray God preſerve the People of* Good Old England, *and Divert his Judgments from us.*

LONDON, Printed for *J. Nutt*, near *Stationers-Hall*, 1699.

FIG 4 The sun eclipsed, from 'The true figure of that great eclipse of the sun', a pamphlet of 1699

By permission of the British Library (719.m.17.(20.))

such foolproof weather predictions, and Poor Robin mocked them for doing so. But the eclipse of 1699 tells us much more about how ordinary people learned of, and were affected by, public mathematical calculations, and how they could go wrong.

The last time a major solar eclipse had been visible from England—in 1652—it had been interpreted (by William Lilly, among others) in apocalyptic terms fitted to the political situation, and it acquired a fearful retrospective reputation as 'Black Monday', when they sky went dark and the heavens showed forth their judgement. On later occasions, such as in 1724, public responses to eclipses took the form of scientific lectures and demonstrations as well as astrological interpretations of their marvellous effects and exaggerated notices in the newspapers that it would be dark enough to see the stars—serious fright was less in evidence. The late 1600s was a period when ideas about these things were changing fast, partly because mathematical techniques for predicting the paths of eclipses were improving rapidly and doing away with some of the exciting uncertainty that had obtained for most of the seventeenth century.

Johannes Kepler had published his geometrical account of the motion of the earth and the planets early in the seventeenth century, placing the moving bodies in elliptical orbits around the sun subject to well-defined rules. The sun lay at a focus of the ellipse (if you draw an ellipse by the gardener's method of putting a loop of string around two fixed stakes and a spade, the stakes lie at the foci), the orbital period was proportional to the three-halves power of the orbit's size, and the line from sun to planet swept out equal areas in equal times. By the second half of the century this was quite widely accepted among learned astronomers as an accurate description of how the planets moved. It was distinctly unhelpful, however, for the purpose of making predictions. The law which stated that the line from sun to planet swept out equal areas in equal times, though relatively straightforward to confirm after the fact, was particularly difficult to work with: knowing the time didn't make it at all easy to work out exactly what angle a planet would have moved through. Newton's laws of motion and gravitation, published in 1687, were, if anything, even harder to work with.

So, astronomical textbooks continued to describe very different geometrical strategies for determining the future position of a given planet. Thomas Streete's 1661 *Astronomia Carolina* ('Caroline' from the Latin form of Charles II's name, Carolus) was fairly typical—indeed, it was one of the most prominent of English astronomy textbooks. Instead of working with equal areas and equal times, he dealt with a planet's variable speed by supposing instead that there was a point somewhere in the solar system from which the planet would appear to move round the heavens at a constant rate. Streete placed that point (call it P) at one focus of the planet's elliptical orbit: the one not occupied by the sun.

This gave him what he called the planet's 'mean motion', to which he added a correction, the 'variation', involving a rather strange geometrical trick. Stretch the planet's elliptical orbit so that it becomes a circle, then project the planet's resulting position back onto the original elliptical orbit along a line joining it with P. Rather amazingly, this gave a reasonable model for the changing speed of the planet over the course of its orbit.

Streete's strategies of calculation went back to a French original, modified by Seth Ward, the Oxford Professor of Astronomy. Although he wasn't a star in the astronomical world himself, Thomas Streete could draw on a constellation of astronomical expertise to refine and justify his strategies of calculation. The particular version of this model which he printed combined published and unpublished refinements by several other English astronomers, and the data he used about the orbits of the different planets went back to Kepler himself, and indeed to the observations made by the great Danish astronomer Tycho Brahe in the days before telescopes.

For a solar eclipse it was necessary to apply this model twice, once for the motion of the earth and once for that of the moon, having first used rougher procedures to determine which of the year's new moons were at all likely to produce solar eclipses. (The moon's orbit is slanted compared with the earth's. Only twice each month does the moon pass through the plane—the ecliptic—in which the earth's orbit around the sun lies. If one of those passages comes close to a new moon, an eclipse is likely.)

Streete worked out the positions of earth and moon at half-hour intervals around the moment of new moon, and then the apparent positions of sun and moon from a particular point on the surface of the rotating earth, in order to work out—for an observer who stood at that point and looked steadily towards the sun—the line along which the moon appeared to move through the sky, and its speed. This could only be an approximation—the moon's motion relative to the sun was not really either straight or at a constant speed—but it allowed Streete to estimate the parameters of the eclipse: the moments of first and last contact between sun and moon, the proportion of the sun that would be covered, and the duration and exact time of the event.

Madly complex though this all sounds, it not only worked tolerably well but was set out in detail in some almanacs (some even advertised their dependence on Streete), apparently so that readers could check the calculations for themselves. In 1699, readers who wished to see detailed calculations could have chosen between half a dozen different almanacs and pamphlets which showed—inevitably with slight differences—not just the predicted time and size of the eclipse, but exactly how they had been determined. The strategies of calculation were all essentially the same, but the writers arrived at slightly different results depending on the location they were calculating for, as well as their own competence. Wing's almanac had just over ten twelfths of the sun eclipsed, Henry Coley a very precise proportion well over eleven twelfths. Neither was afraid of precision: Wing asserted that the eclipse would begin at exactly thirty-three minutes and sixteen seconds after eight in the morning, and last until ten fifty-eight and eleven seconds, Coley that eleven digits (twelfths) of the sun would be eclipsed, plus forty-eight sixtieths of a digit and thirty-nine sixtieths of a sixtieth.

Meanwhile, purchasers of less technically-minded almanacs received a much more bald report:

> The *beginning* thereof will be about 45 *minutes* past eight; the *middle*, or *greatest*, *Obscuration* about 41 *minutes* past nine; and the end about 35 *minutes* past ten *of the Clock* in the said *forenoon*.

William Andrews, who was fairly specific about the time the eclipse would happen, gave an estimate of the magnitude of the event as simply: 'the *Suns Body* will be somewhat above ten *parts darkened*.' He provided a simple picture to show what a ten-twelfths eclipse might look like, but he made no claim to have calculated these things for himself.

More common still were even vaguer reports: little more than a note that the eclipse was going to happen. Isaac Abendana used almost exactly the same words as a dozen other almanac writers:

> an Eclipse of the sun on *September* the 13*th*. about eleven of the clock before noon ... it will be a great and visible Eclipse, and ten Digits of the Sun's body will be darkned.

So whoever you were, and whatever level of detail you wished for about the eclipse, there was an almanac for you. Most had astrological predictions as well as astronomical data, and all were, as always, tailored to the political and other preferences of the targeted reader. Those with a taste for classical learning could read summaries of ancient writings about eclipses, and those interested in recent history were regaled with stories of the remarkable effects of the 1652 eclipse.

The very keen also had an opportunity to buy one of the eclipse-related pamphlets which circulated during the fortnight or so before the big day. The spectacular diagram in Figure 4 appeared in such a pamphlet—it certainly looked both learned and convincing. Its detailed information gave the superficial impression that the author was a real expert, in a position to impart wholly reliable information. The top part of the diagram was very much the kind of thing envisaged in Streete's instructions for calculating the time of an eclipse—the moon's track relative to the sun—although Streete made no mention of the moon's prominent nose which is such a feature here. The bottom part looked as if it could be a detailed and accurate representation of the moon's shadow at a particular moment, but in fact it seems to have been no more than a vague guess. In particular, the author seemed to state—wrongly—that the central part of the shadow, the path of total eclipse, would pass over London.

A couple of other pamphlets produced around the same time stated that eleven-twelfths of the sun would be covered from London. That was at the very high end of what could be found in the almanacs, and it seems these last-minute authors were taking the most sensational figures they could find, and in some cases exaggerating them still further. The resulting pamphlets would sell, by virtue of being sensational, but it was a short-term strategy, since an author's reputation would probably never recover if the eclipse didn't live up to predictions.

All of this printed discussion of the eclipse would have been magnified by discussions, perhaps in taverns and coffee shops, a route by which even those who didn't or couldn't read an almanac or an eclipse pamphlet could learn something about what was to be expected on 13 September. You could hardly have avoided knowing the eclipse was coming.

A really interested person could have seen a whole range of different predictions about the eclipse, the results of a complex process of mutual copying, elaboration, simplification, and garbling. At each stage, people received information which they couldn't check, and passed it on. Almanac writers carried out Streete's calculations without being able to check whether his strategies of prediction were good or bad. Ordinary people read about the eclipse in dubious pamphlets and told their friends about it without having any way of knowing whether the predictions they'd read were accurate or not. If the astronomical information involved went back to Kepler and Tycho Brahe, the pamphlets printed at the last minute owed little to anything other than sensationalism.

Given this complex and confused trade in eclipse lore around England, the non-specialist had little chance of separating fact from wild speculation, or of learning anything about the different degrees of confidence that might apply to different predictions. That there would be a solar eclipse was in no real doubt: the time of its maximum extent had to fall within a few minutes of the moment of full moon, which could be calculated accurately without too much trouble. The width of the moon's shadow might be found by quite straightforward geometry with reasonable accuracy, by those who knew how. But just *where* the moon's shadow would fall was a hard question even for the most knowledgeable specialists.

Mathematicians had no well-developed language for speaking about error or approximation, and the writers of much of the eclipse material had no interest in giving their readers any sense of their own uncertainty. But with most almanacs predicting an eclipse of about ten twelfths of the sun, the well-respected Coley a very precise figure around 98.4 per cent, and some pamphlets a total eclipse, the person in the street would have been reasonable to conclude not just that the experts disagreed but that something very spectacular was expected.

Sadly, though, it was all a game of whispers. The bulk of the almanacs were about right with their figure of ten twelfths, and the belief that the eclipse would be spectacular and obvious was largely the product of rumour. It was also a result of the difficulty of the calculations involved, and their sensitivity to small errors: in a solar eclipse the sun's shadow may be only a few tens of miles wide and move at speeds in the region of a kilometre per second, and some writers evidently did the calculations conscientiously but got them wrong. Either way, an unspectacular event was transformed into a major public phenomenon, a detail in the astronomical calendar turned into something that would bring people onto the streets and induce them to pay large sums of money for the best viewing places.

After the eclipse the London newspapers tried to excuse themselves as best they could. Londoners may have been made to look like fools, but not as badly as some.

> All the North of *England* was under the greatest consternation imaginable upon it; and fear'd the Eclipse more than they wou'd have done the Landing of the *French* three Years ago. In some Villages they lay a-bed all day, the Barbers refus'd to Trim their Customers; the Drovers wou'd not drive their Cattle, for fear the Eclipse shou'd surprize them.

Sporting events—horse races and fencing matches—were cancelled, as were 'Weddings without number'. 'In short, it put a stop to all manner of Trade, and business'. Contemporaries estimated the damage to trade amounted to £20,000 in the north of England alone. What was advertised as a public spectacle had now become a public nuisance.

The trade in mathematical information around early modern England was a complex one, and it didn't necessarily work very well. When things went wrong, though, it was the reputation of mathematics that suffered. The hapless Londoners who spent all morning waiting for the sky to go dark, while the total eclipse made its way through the northern tip of Scotland, Denmark, Poland, Ukraine, Turkey and Iran, probably chalked the experience up to the well-known deceptiveness of mathematics.

> My youth and beauty still remain;
> Yet, you see, I have lost the swain.
> Ah! my girl, the thing's too certain;
> The pangs he felt were for my fortune.
> Why—five and forty—thousand—pound
> Had given the Great Mogul a wound!

The mighty Czar, had he been living,
Had thought the present worth receiving.
But that delightful South Sea scheme—
That charming, warming, golden dream,
Which made so many fools and knaves,
And left so many well-bred slaves—
Fell to the depths from whence it came,
And quenched at once his towering flame.

Another piece of mathematical deception. As 'Cloe' wrote to 'Aminta' in this mocking bit of pastoral published by the Canterbury poet Sarah Dixon two decades later, the year 1720 was a bad one for personal fortunes, in both senses of the word. By 1722, Poor Robin's list of notable dates included 'the late prodigious high Tide in the great *South Sea,* which was nine or ten times as high as any had been known there for several Years', and asked the reader 'Whether that high Tide did not wash down abundance of Gentlemen's Houses about their Ears, drown their Estates, and leave them poor and miserable'. He hinted that there were consequences further down the social scale too, with the date when 'a Man was arrested for Money that he owed to a Taylor in *Spittlefields.*'

The story of the South Sea Bubble has often been told. The stock market was already a source of anxiety and resentment to British people in the early eighteenth century. Its emergence in the new political conditions after the revolution of 1688, together with the creation of the Bank of England and the National Debt, has been called the 'financial revolution', a radical change in financial behaviour ultimately founded on the improved solvency of the state. King William and his successors the Hanoverians were able to offer better security to lenders than the Stuarts had, and therefore to raise much more borrowed income: public debt could now be guaranteed by Parliament.

One manifestation of the financial revolution was that the number of joint-stock companies rose extremely rapidly after 1688.

They accounted for a ninth of the country's personal property by 1695, and there was no tax on the yields from stock. By the early eighteenth century London had become a major centre of financial activity. It had commercial colleges, a national bank to loan money to the government, and financial pamphlets by the hundreds. But although there was plenty of economic theory, the mysteriousness of these ways of making money led inevitably to resentment by some.

The South Sea Company's origin in 1711 consisted of a restructuring of the national debt: the exchange of government debt for shares in the new company. Put another way, it effectively incorporated certain holders of public debt—millions of pounds' worth of it—as a company. By the middle of the decade the company had two or three thousand shareholders, and from 1717 George I was its formal head. The company existed, in principle, to exploit a monopoly on British trade in the South Atlantic, but for much of the 1710s its actual South Sea trade was non-existent, and the boom and bust that made it notorious was a purely financial affair.

In 1719 the directors of the company negotiated an Act of Parliament in which a substantial part of the remaining national debt, most of which took the form of annuities, would also be converted into South Sea Company stock. The Company's dividend record made the deal acceptable, even desirable for holders of government debt, who would have the chance to profit further if the stock rose in price. Late in 1719 peace with Spain was in the air, making the Company's trading prospects appear bright, and the Company would enormously increase in notional value when the new stock (plus a great deal of extra stock authorized under the deal) was issued. Parliament, meanwhile, would be relieved of servicing a very substantial amount of debt, in exchange for a relatively modest annual payment to the South Sea Company. The deal was improved still further by the promise of a large cash gift from the Company to the nation once the exchange of debt for stock was complete—bidding against the Bank of England, which would have liked such a deal for itself but ultimately offered less impressive terms, resulted in the size of that gift standing at seven and a half million pounds.

That figure gives an idea of the sums involved. The total value of the debt to be converted was estimated at around thirty-one million, and the government's total payments to the Company were nearly two million a year.

The Company's directors now promoted it by means of a two-part cycle. First came the sale of new Company stock (prematurely, since the exchange that was supposed to justify the creation of the new stock had not yet taken place) at high prices and on attractive buy-now-pay-later terms. Second came the release of the cash thus gained (and, on one occasion, of a million pounds in Exchequer bonds, loaned to the Company for the purpose) back into the market in the form of loans on the security of the stock itself. The scheme created credit, but credit which was intended to remain within the Company's control.

There were three rounds of sales and three of loans, and the scheme was further assisted by the transfer of very large amounts of fictitious stock to 'friends' of the Company, and by action by the House of Commons to quash rival companies. The skilful timing of the releases of cash—and of the above-mentioned Exchequer bonds—boosted the Company's credibility and sent its stock prices soaring, making the conversion to stock attractive to holders of government debt. The conversion itself—at rates determined by the Company—of course created more new stock, giving the Company enormous paper profits and increasing its credibility still further.

Disaster was not inevitable: the obligations the Company took on were by no means absurd in relation to the expected total value of its stock after the conversion. Yet repeated rounds of money subscription and loan—the inflation of credit—raised the price of stock to dangerous levels.

Some stock-holders (including Sir Isaac Newton) sold quite early in 1720, taking modest profits (Newton is sometimes said to have re-entered the market later in the year and made heavy losses, but it seems this may be a myth). Guy's Hospital in London was built on the profits from cautious sale. Another gainer was Robert Smith, Plumian Professor of Astronomy and Experimental Philosophy at Cambridge, who donated three and a half thousand pounds' worth of South Sea Stock to found the Smith's Prizes in mathematics at his university. Others used their gains to buy estates, while others again, like 'Cloe' above, built castles in the air upon paper gains not yet realized.

Throughout the first half of the year people remained keen to acquire South Sea stock. A million's worth was said to have been subscribed for in just an hour when the first sale opened, at nine in the morning on 14 April. By late spring the demand had spread all over Europe and, in the run up to the conversion of government debt, quotations of the price of South Sea stock were printed in local newspapers across England. The market reached such a frenzied state that, it is said, stock was selling for different prices at the two ends of Garraway's coffee house.

But by the end of June the cycle of loans and rising stock prices was flagging, and the Company had lent much more than it had received in cash. By July, stock was changing hands for nearly ten times its face value. In mid-August the holders of the remaining pieces of national debt were offered their exchange, but the offer the Company could afford to make was not one they were happy with. Rival companies were also causing problems for the Company, despite legal proceedings against them by the Treasury. But optimists were still confident of profit, and South Sea stock was still saleable at very high prices.

There was a fourth money issue in August, restricted to existing shareholders. But this did not raise enough cash to cover immediate needs, and disaster was now at hand. Promises of guaranteed dividends were announced in an attempt to maintain the high price of stock, but this failed to impress—it amounted to an offer of modest yields to buyers who had hoped for large capital gains. Meanwhile, commercial concessions from Spain had yet to materialize, leaving the Company still, albeit temporarily, unable to make any profits from actual South Sea trade. Contemporaries tried, and failed, to calculate the 'real value' of South Sea stock. The directors of the Company received death threats.

In early September the price of stock collapsed by a half in a few days. By the end of the month it was down to a less than a fifth of its peak. The South Sea's sister company, the 'Sword Blade' Bank, was unable to meet its cash obligations. The later subscriptions had to be re-valued, 'writing down' the total value of the Company's stock and surrendering tens of millions of pounds in stock price, and tens of millions more in future instalments of payment on the subscriptions already taken.

The subsequent furore, investigation, and partial cover-up have become the stuff of legend. Those who had sold early and gained had little reason to advertise the fact. Those who had lost paper fortunes or real fortunes had every reason to seek redress, and little left to lose by doing so. The real scale of the losses remains hard to assess, and the affair is clouded by the hysteria of the period. One contemporary judged that two thirds of the whole credit in the country had become waste paper, another that South Sea transactions had been made by lunatics and should therefore be void. One resentful contemporary blamed a deceitful arithmetic:

> one added to one, by any rules of vulgar arithmetic, will never make three and a half; consequently, all the fictitious value must be a loss to some persons or other, first or last. The only way to prevent it to oneself must be to sell out betimes, and so let the Devil take the hindmost.

And here's Anne Finch, Countess of Winchilsea, in her 'Song on the South Sea':

Bright jewels, polished once to deck
The fair one's rising breast,
Or sparkle round her ivory neck,
Lie pawned in iron chest.

Furthermore, assessment of the South Sea Bubble was and is clouded—to an extent—by the fact that much of the contemporary commentary was written by people deeply suspicious both of the new-fangled stock market, with its allegedly immoral logic of 'devil take the hindmost' and 'selling the bear's skin before the bear was dead', and also of its potential to make foreigners, Jews, Catholics, women, and other undesirable groups into *nouveaux riches*. Thus, crude and basic prejudices at every level of society were engaged—much contemporary commentary consists, in effect, of loaded warnings to maintain the social status quo. Figure 5 shows one contemporary view of the matter: a playing card illustrating a man who invested his servants' wages in South Sea Stock without their knowledge and gave them the profits. Benevolent the 'Good Old Worthy' may have been, but enriching one's servants was not a universally popular thing to do.

Poor Robin, speaking on the whole to those who had not bought stock but who might have been affected indirectly by the eclipse of credit, was scornfully unsympathetic. Despite his different perspective he was another fan of the status quo:

> If a Man makes me Pay Five Pound for Five and Twenty Shillings, that is a Knavish mistake. If I let him cheat me again that cheated me once before, that is a Foolish Mistake. If I pay Money in my own Wrong, only upon the word of a stranger, that is a careless mistake.

So it's worth dwelling on just what went wrong. In the judgement of the best recent historian of the Bubble, Helen Paul, despite the legends there was no 'gambling mania'. Most investors were not 'playing' in the same way that some Georgians would play cards for money, knowingly taking risks for the fun of it. The South Sea

FIG 5 A 'South Sea Bubble' playing card

Private collection © Fine Art Images/SuperStock

Company was no sham, and there were sound reasons to invest in it. It had a guaranteed income from the state, it had unique access to the valuable trade importing slaves to Spanish South America—admittedly a trade temporarily halted by Spain's declaration of war—and there seemed the realistic possibility that the Company might gain a colony of its own in South America. Thus, its prospects were solid enough to make investment a rational choice. Once the enthusiastic market for South Sea stock had been established, it was rational to predict a price rise and thus to invest.

The Bubble was therefore what economists call a 'rational bubble': one where price rises are driven by individuals acting rationally but with limited information. Some, of course, were influenced by fads and false news, as in any market, but that was not the determining factor in the events of 1720.

The collapse affected not just those who had invested, since through them a wider world of trade, both in luxury goods and even in staples such as coal, became depressed. But the real human cost of the South Sea scheme was not to countesses who pawned their jewels or to middle-class investors who made and lost paper fortunes. It was to the slaves—more than thirty thousand of them, largely invisible to Georgian contemporaries—the South Sea Company imported to South America during the longer course of its existence. They are conspicuous by their absence from contemporary comment.

Among its other consequences, the collapse ushered in the first of the great Georgian Prime Ministers, Robert Walpole. In the chaos and acrimony of the parliamentary investigation he did much to devise a solution to the financial mess, and he was an active man in the prosecution of the guilty parties. In that solution many of the Company's loans—and the cash gift it had originally promised the Exchequer—were cancelled, and its stock allocated in varying amounts to its proprietors, those who had subscribed money, and those who had exchanged annuities and other government paper. The finding of the parliamentary investigation was that, in the words of historian John Carswell, 'in lending money on the security of its own stock the directors had been guilty of a breach of trust and ought, out of their private estates, to make good consequent loss to the Company'. A bill putting this into effect passed through Parliament within a few weeks. In the end the size of each director's contribution to making good the losses was individually assessed on the basis of the evidence, together with an inventory of his current personal wealth. If, as has sometimes been alleged, the directors of the South Sea Company had gambled on ensnaring so many highly placed people as to place them beyond reprisal, they had failed. Poor Robin approved of such retribution, and crowed over the new poverty of 'Stock-Jobbers', 'Thanks to a good Government'.

The retribution was not complete. The responsibility for the Company, if not its failure, has been judged by most to have largely rested with the company's cashier, Robert Knight. He fled the country, intending to take refuge in Germany with John Law, who was behind the inflation of a recent and broadly comparable bubble in France. In fact he ended up in the Austrian Netherlands, from which he could not be extradited. The embarrassment of not being able to lay hands on Knight threatened not just the government but the whole Hanoverian settlement, and negotiations with Austria had to be carefully 'screened' from the British public gaze. Walpole, thus 'screenmaster' from the start of his reign, would ultimately become a controversial and even a hated figure—as it happens, he was universally known as 'Cock Robin', his reign (1721–42) as the 'Robinocracy'. (A prominent politician and man of twists and turns of the previous generation, Robert Harley—Earl of Oxford and Prime Minister in all but name from 1710 to 1714—had been known as 'Robin the trickster'. A connection with Poor Robin is, sadly, unlikely.) Knight's escape, meanwhile, was eventually arranged by Britain and Austria, and he lived on in France for two decades.

Whether individual paybacks could be arranged or not, the paper fortunes of mid-1720 were gone. The loss of a private fortune to the oblivion of a share-price crash must have struck many of the victims (and their dependents) as an inexplicable failure much like the failure of the eclipse in 1699. Predictions had been made and promises held out by those who claimed technical expertise in these essentially mathematical matters, but in the fullness of time the promises proved to be empty and the expertise powerless to protect individuals.

So, reasonably or unreasonably, there were implications for the reputation of economic expertise. Carswell writes that by December, 'People had had enough of expertise'; the writer and politician Richard Steele spoke scornfully of 'cyphering cits' (that is, calculating citizens). It is not clear how far the famous 'Bubble Act' really limited subsequent economic growth, but the Bubble, the bursting of the Bubble, and the mess left afterwards did provide an enduring language with which to talk about the socially suspicious activity of stock-jobbing and the dangers of numerical cleverness.

One of those whose reputation had most impact on the reputation of mathematical learning was the South Sea Company's accountant, John Grigsby. Grigsby had kept a coffee house in Threadneedle Street, behind the Royal Exchange, in the late seventeenth century. He became a licensed stockbroker under the legislation of 1697 which regulated that profession for the first time. In 1711 he became the accountant for the South Sea Company, where his main function was to register changes in ownership, and to make new issues of stock. One of the most powerful of directors of the South Sea Company, the 'decemvirs', as the contemporary phrase had it, he was not only privy to its policy decisions but also directly responsible for the turnover of many millions of pounds worth of stock.

Carswell writes that:

> The exotic commodity in which he had once dealt, the sinister profession he had later adopted and his gnome-like appearance combined to give him a vague reputation for dabbling in black magic and the nickname of 'The Necromancer'.

A contemporary said more:

> Nor was it absurdly imagin'd of the town, to take a man for a Negromancer, Conjurer, or what you please more artful, who cou'd bring his horses to eat gold, when they did not like hay; and from a grinder of Coffee so to order his affairs, that a noble Duke and a Marquis thought it an honor to support him under each arm, being crippl'd with the gout.

Grigsby was by no means personally blameless, harsh though this abuse may sound. When a dividend was issued in the form of a bonus issue of stock, for example, he was responsible for buying up options on this new stock on behalf of certain of the directors. Later he conspired with Blunt, one of those directors, to exchange Blunt's stock for cash shortly before the crash. He personally profited handsomely from the Bubble: he bought a country house and was said to be worth £50,000 in mid-1720. At the Company's governor's birthday gala he was helped out of his coach by the Duke of Marlborough: a most spectacular—and most undesirable—*nouveau riche*.

Early in 1721 Grigsby was examined by both the investigating committee and the House of Lords, and with the other directors he was ordered to make good the Company's losses out of his own wealth. In a subsequent stage of the investigation, after Knight's flight, Grigsby was arrested and his papers seized from South Sea House. By this time so infirm that he was unable to hand in his inventory in person, Grigsby was nonetheless dealt with very harshly and allowed to retain only £2000 of the more than £60,000 he now owned. He died in 1722 and was therefore spared, perhaps, some of the long-term effects of the opprobrium in which he was held.

Grigsby was someone whose only claim to fame was the use he had made of his numerate skills, and that fame was a very dismal one indeed. It is not clear quite how brightly he shone in the public memory—although his coffee house remained in business to keep his name in view. It may be no more than a coincidence, but in John O'Keefe's comedy *The World in a Village* of 1793 a character shared his name: Dr Grigsby, a former barber, now on the make. Another character remarked 'The fellow has made money out of people's folly, and now don't know how to behave himself.'

The first decades of the eighteenth century were a particularly interesting time for mathematics. Discoveries had been made during the previous century that would change the subject—and the world—irreversibly: Kepler's and Newton's laws of planetary motion; the development of algebra and of calculus. As we'll see in the later chapters of this book, there was real utility, both public and private, to mathematics, and real optimism about what it could achieve for the individual and for society.

These decades were also an interesting time for the reputation of mathematics. The examples of the eclipse of 1699 and the Bubble of 1720 give an idea of the kinds of things that could go wrong for mathematics and its reputation. But there were wider contexts and longer cultural memories at work, too. A long tradition took mathematics to be a form of conjuring—it was fuelled in the seventeenth century by the astrologers and their work, and the abuse thrown at John Grigsby ('necromancer') shows that it was still not quite dead. Indeed, it would not completely vanish until the end of the eighteenth century, if then.

There were plenty of reasons to suspect mathematics, for those who wished to do so. Arithmetic was involved in the making and breaking of codes, for example, while geometry was responsible in part for the illusions of the theatrical (and the magical) stage.

'Strange and mysterious Algebra' had already something of its modern reputation for incomprehensibility. Isaac Newton himself wrote memorably of how he had written the third book of his *Mathematical Principles of Natural Philosophy* 'in the mathematical way' specifically so as to limit its readers to those 'who had first made themselves masters of the principles established in the preceding Books'. So if numeracy could be a way to avoid being 'cozened', it could also be a tool to conceal and even deceive. As Poor Robin continued to remind his readers, any specialist language could become a way to cloud the issue, whatever the issue may be—and mathematics was no exception.

Public manifestations of numerate expertise had, as we have seen, an embarrassing tendency to belie the new confidence and power of mathematics, producing false astronomical predictions and failures of public finance that could have substantial—even devastating—results for individuals. Blunders are memorable, successes less so, and the memorable blunders did not cease. As late as 1751 a report in the *Philosophical Transactions of the Royal Society* concerning a recent occultation of Venus by the moon lamented that 'many had gotten a notion from the almanac-makers' that—incorrectly—the event could not feasibly be observed. Financial mismanagement was a perennial theme of political discussion throughout the century, and the consequences could stretch right across society.

> It is the Opinion of this Committee, That 5s. for each Man *per* Month, of Twenty-eight Days, in a time of War, is an indifferent Estimate for the Charge of the Wear and Tear of the Ships of War, of the Wages and Victuals for the Men, and of the Charge of the Ordnance and Amunition, and all other incident Charges [...]
> the Committee could not examine what Number of Men were required for the Sixty-five Ships of War, and Eight Fire Ships; there being no List of the Quality of the said Ships, as mentioned in the Estimate; and so cannot assert what the total Charge of the said Ships amounts to: Neither could the Committee examine, what Number of Men were necessary...

> But if the Sixty-five Ships, and Eight Fire Ships, have 17,155 Men employed on them, the same, at 4*l.* 5*s.* *per* Man, will amount to 72,908*l.* 15*s.* *per* Month; and, for a Year of 13 Months, to 947,813*l.* 15*s*...
> And the Wages is not payable till the End of the Year.

So ran parliament's deliberations (in part) in 1689 concerning the 'navy estimates': estimates of how much the Navy could reasonably be expected to cost during the coming year. In its printed form the discussion went on for some pages: it was a lengthy and detailed debate, with plenty of numerical information provided and calculations carried out. It might give—indeed, it was perhaps meant to give—the unwary reader the impression that parliament interested itself deeply in these details and the computations by which they were arrived at, and that it took pains to get the figures right. Crown and parliament repeatedly authorized the printing of broadsheets or pamphlets setting out in detail where this particular parcel of public money was going, as well as what it was buying the country in terms of the defence of its own shipping and the capture of its enemies', reinforcing the impression of detailed financial control of this most important institution.

On closer inspection, though, the pattern was one of chronic underfunding, and of an ingrained optimism of what might be achieved with less money than was really needed. For the Navy as a whole the consequences were serious: over time dockyards became unwilling to take on work for the crown, for example, unless legislation forced them to do so, because in practice it was unlikely they would ever be paid in full. Ships deteriorated, new ones were built slowly, and in the words of Nicholas Rodger, a historian of the British Navy:

> The Navy Estimates allowed 19s or 20s a man a lunar month for victualling, but the real cost of four weeks' naval victuals was about 23s, rising to 30s in years of bad harvests, and the number of men actually serving in the 1690s was sometimes as much as 40 per cent higher than that nominally voted. Prisoners of war were invariably, and troops

transported overseas often, fed from naval stores without any additional vote to cover them.

The problems persisted for decades, during a period when Britain was repeatedly involved in conflicts overseas and which saw, during the early eighteenth century, Britain's emergence as a major international power. That status, and the European obligations of the house of Hanover which held the throne from 1714 onwards, made the condition of the Navy more and more of a concern for the country and (foreign policy being the crown's business) the monarch.

For individuals, things could be equally serious. One of the heaviest burdens fell on ordinary sailors, who were customarily paid their wages when their ship returned to England. That required the Navy Board, which administered such things, to put aside a substantial sum of money with which to pay those wages when required. In practice, the temptation to spend that money on other things proved overwhelming, and the Board fell again and again. When the sailors did return home they were met with empty coffers.

The Navy Board attempted to manage the situation by instituting an ingenious system in which pay tickets were to be redeemed not at a specified future date and certainly not when their holders chose, but strictly in the order in which they were issued. When a particular group of tickets was coming up for payment, a notice could be placed in the newspapers announcing the fact.

By this novel system the Navy Board raised what amounted to a public loan without either paying interest or specifying a date of repayment. It goes without saying that it was resented terribly by ordinary sailors, who found themselves forced to participate in the scheme with money which they neither had nor could afford to lose. In order to obtain cash for immediate needs, they were obliged to sell their tickets to third parties at a discount, or to get involved in more complex and doubtful arrangements such as loans raised upon the security of the tickets. Naturally, since the sailors were in a weak position, the value of wage tickets for such purposes was low. Equally, a common sailor did not have the appropriate skills to establish whether a particular price was fair or a particular offer reasonable, making him easy prey for the unscrupulous: after all, what did constitute a fair price for a wage ticket or for a time-limited loan raised upon it, with no guarantee that the ticket itself would become payable within the duration of the loan? In 1700 a group of sailors had a broadsheet printed complaining about the situation, and setting out some of the things that could go wrong.

> Others, when Tickets have been sold at a very low and under rate, by poor Sea-men, have taken Bonds from such Sea-men for the payment of so much Money at a certain day, or at the payment of their Wages by the King, and have put the same in suit, and recovered thereupon, pretending they never received any such Money, and sometimes that the Bonds were absolute, and had no reference to the assignment of the Tickets or Wages, altho the said Bonds were given only as a Collateral security ...
>
> Letters of Attorney, Assignments, or Bills of Sale of Seamens Wages have been often antidated, and the persons to whom the same have been made, have received the Wages upon such antidated power, and so deceived others, who have had an honest right to their Wages by Assignments lawfully executed.

The situations described seem horribly intricate, and this plaintive document—which would have been pasted up on the walls of London—illustrates what the consequences could be for ordinary people of Parliament's optimistic naval estimates. Such complaints as this one were frequent, and sometimes they listed the names of individual sailors and the moneys, or simply the years of pay, owed to them. Even such blatant attempts to embarrass the Navy Board, and through it the government, were ineffective. The money which should have been paid to the sailors was, after all, long gone.

So arithmetic carried out by the state could be just as fallible as that carried out by individual speculators or the writers of popular astronomical works. The system of seamen's tickets tried to shift the consequences—or indeed the burden of arithmetical competence—onto private individuals. When the individual could not bear it, things could go wrong very rapidly.

Literature on the subject of debtor's prison from the eighteenth century sometimes has a mawkish tendency ('My children's cries still vibrate in mine ears'), but people of many ranks were hit by the disaster of imprisonment for debt which could, notoriously, involve the victim in a spiral of unpayable obligations within the jail.

You didn't, of course, have to be numerically incompetent for it to happen to you. Such successful naval figures as Edward Dummer, who ran packet services to Lisbon and the West Indies during the early 1700s, and Charles Clerke, one of Captain Cook's commanders, would find themselves in prison for debt: Dummer died there (in the Fleet) in 1713. One of Handel's trumpeters, John Baptist Grano, ended up in the Marshalsea Prison for over a year in 1728–9, and left a moving diary of his experiences. This was during a period when deaths could number ten a day in the Marshalsea, of which three were from starvation. The consequences of computational failure could be serious indeed. It's no wonder that the reputation of mathematics continued to be two-sided.

Its appearances in fiction reflected that. Here's a piece of truly dangerous mathematics:

> [H]ad I been a Lilliputian born... to have brought those Figures to light, would have drawn on me some very severe Punishment; but as I was a Stranger, and had been guilty only through Ignorance, 'twas probable I might obtain Pardon from the Emperor, if he should happen to know it; however, he advis'd me to conceal what I had done, and erase the Figures, if by any means I could, so as they might not be seen by any that came to visit me.

This narrator had discovered some geometrical diagrams concealed beneath a layer of plaster on a wall and, as he explained in this fantasy from 1727 set in the 'court of Lilliput', the penalty to which they made him liable was death. Duly pardoned as an ignorant foreigner, he learned that:

> this Painting...was done by the greatest Artist of his Time, from a Draught given him by the first and perhaps the greatest Philosopher, Mathematician, and Geographer that ever the World produc'd.

The fashion for mathematics which ensued from that genius' fame, and the dogmatism of its 'imaginary Proficients' had disturbed the peace:

> Every one had a particular Set of Followers, who appear'd so well convinced of the Truth of what they profess'd, that they declared themselves ready to endure Martyrdom for the Conviction of the rest.

Meanwhile the fad 'took our Youth from the more useful Studies of War, Politicks, and Mechanism', and at last the Emperor 'made an Edict, strictly prohibiting the use of Mathematicks for the future, except in such Branches of it as were necessary for Navigation, or for Weight and Measure', on pain of a large fine, or in lieu of payment death.

> All the Books of Argument relating to this Science were immediately burn'd, all the Paintings of it demolished, or plaister'd over...and the same Punishment allotted for any one who should conceal the one, or by any means preserve the other, as for him who should be guilty either by writing or painting a new one of the same kind.

The narrator approved of these measures against mathematical studies, 'thinking it very prudent in the Government of Lilliput to suppress them, when they began to encroach on the practical and more useful Business of Mankind'.

The 'greatest philosopher' recalls Newton almost irresistibly, making this apparently, at least in part, a comment on the Royal Society and its Newtonians (like *Gulliver's Travels* itself, the inspiration for this memoir of Lilliput). This remarkably detailed discussion of the rise and fall of mathematics represented an extreme case, but there was a tradition for this kind of thing in utopian writing: mathematicians who could not agree on the longitude and hence caused one disaster after another; geometers whose disputes erupted into civil war and had to be banished to a small island.

> [T]he philosophers and geometricians have been forc'd to inhabit a dry, barren country, in which nothing is found but bitter fruits, and is besides choak'd up with briars and brambles. 'Tis here the geometricians spend the day in drawing figures on the sand, and in demonstrating clearly to one another, that one and one make two; and the night in observing the cœlestial motions . . . they meditate so intensely on curves, obtuse angles, trapeziums, and other mathematical figures, that one wou'd conclude their minds were moulded into those shapes.

Yet at the same time, understandably, there were writers who were working hard to improve the reputation of mathematics:.

> Q. Having lately read in one of the Books of Dee's Euclid something concerning perfect Numbers, and it not being in my Capacity (being but a young Student) to comprehend the true Notion thereof; I beg your Assistance in it, so far as to satisfy what they are, and by what Means I may find any one of them out, for I find them to be of great Use to me, and in so doing you'll highly oblige your Friend, etc.

This was an anonymous reader, writing in to the *Athenian Gazette or Casuistical Mercury*. The *Mercury* was a publishing phenomenon of the 1690s, sometimes described as a 'poor man's *Philosophical Transactions*'. It promised to 'resolv[e] weekly all the most nice and curious Questions propos'd by the Ingenious', and it consisted of readers' questions together with answers provided by a team of four authors. The tendency—when the questions were not provided by the editorial team itself—was for questions to try to probe the writers' stance, particularly on theological matters, and something of that is evident here.

The question about 'perfect' numbers, particularly coupled with the name of John Dee, the great Elizabethan mathematician still notorious for his alleged conversations with angels, invited an answer of a numerological character. In fact it got this:

> A. A perfect Number is that which is equal to all its aliquot Parts added together; according to this Definition, 6 is a perfect Number, because if you take its aliquot Parts, which are 1, 2, 3, their Sum will be equal to 6; again, 28 is a perfect Number, because its aliquot Parts, 1, 2, 4, 7, 14, added together make 28. Now if you will find as many of 'em as you please, take the following Progression: 1, 2, 4, 8, 16, 32, etc, which it is easy to continue in doubling every last Term. Choose in this Progression any one Term; substract Unity from it: if the Remainder is a prime Number, multiply this Remainder by the Term immediately preceding. The Product will be a perfect Number. But if the Remainder is no prime Number you must choose another Term.

This was written by Richard Sault, the mathematical member of the Athenian team. The others were John Dunton, who was Sault's brother in law, John Norris, a moral philosopher, and Samuel Wesley, the father of John Wesley. The group described itself as the 'Athenian Academy' and the popularity of the publication seems to have done well for its authors. Sault later ran a mathematical boarding school in London and taught algebra on a private basis, and he published a short work on algebra as well as translations of historical and theological works from French and Latin.

His other mathematical answers in the *Mercury* gravely solved problems in algebra, worked through textbook problems that were giving readers difficulty, or tackled issues in the philosophy of mathematics such as whether unity should be considered a number. He was never a fellow of the Royal Society, but he was acquainted with a number of elite mathematicians and scientists, and made it clear that he knew their works well, particularly Newton's. His aspirations for mathematics were obvious: in his hands it was clear, powerful, general, and nothing to do with numerology. On another occasion he wrote in the *Mercury* that 'for mathematicians to advance anything that won't bear a demonstration, is worse than doing nothing at all'. When he was asked to recommend a list of mathematical books, he named exclusively élite works in Latin, conspicuously failing to mention that there might be anything worth reading in the English tradition of practical mathematics from which he himself sprang.

His distance from the numerical quackery satirized by Poor Robin and others was obvious from his words about perfect numbers, but he hammered the point home:

> This is a very easy Rule, and we expect our Querist will be pleased with it: but we have something of far greater Consequence to him, which is, that we guessing by his Question that he is apt to attribute some Virtue to perfect Numbers, (or else why should he think they can be of great Use to him?): and [the] Doctrine of attributing Virtue to Numbers being a pure Chimera of Cabalistic Sprits, we advise him to employ his Time better, than in such a vain fruitless Contemplation.

By protesting more than he needed to, Sault gave an impression of quite personal anxiety—the possible association of mathematics with non-existent chimeras seems to have been a matter of real concern to him. Elsewhere, introducing a catalogue of the achievements of mechanical and mathematical ingenuity (marvellous clocks, flying wooden eagles, and the like), he wrote ruefully that:

> We read of many Persons, who in this Study have trod so near upon the heels of Nature, and dived into things so far above the Apprehension of the Vulgar, that they have been believed to be Necromancers, Magicians, etc., and what they have done to be unlawful, and performed by Conjuration and Witchcraft, although the fault lay in the People's Ignorance, not in their Studies.

The satirical blows of Poor Robin and his colleagues, and their association of mathematics with puritan astrology, numerology, and even magic, were not forgotten: their legacy still haunted mathematics in the early eighteenth century. Together with such public disasters as the South Sea Bubble, the failed eclipse of 1699, and the perennial mismanagement of public finance, they created a difficult situation for those, like Richard Sault, who would represent mathematics as benign, useful, and desirable in the Georgian world.

Chapter 3

'Fitted to the meanest capacity'

LEARNING IT

On 13 June 1726, in the village of Linton in Kent, twelve-year-old Isaac Hatch began work in his arithmetic exercise book. It was a ready-bound book of blank paper about six inches by eight, decorated on the front cover with an engraving of a countryman and his dog. 'Isaac Hatch His Book' he wrote, and turned the page.

Hatch's penmanship was a little uncertain. His teacher did most of the writing in the first few pages of his arithmetic book. 'Numeration', the teacher wrote, 'is the first part in Arithmetick and teaches to Read rite or express any Sum or Number of Figures', and set out a chart showing how numbers with up to nine figures should be read. The chart mimicked the ones found at the start of many printed arithmetic primers.

Arithmetic, Hatch learned, had five parts, of which 'numeration' was the first. The others were the operations of addition, 'substraction', multiplication, and division. Each was subdivided according to whether it concerned money, weight, or measure, and separate pages were devoted, for instance, to operations on the different systems of weights: Troy or Avoirdupois. When a new type of operation was introduced, the teacher wrote out a definition or an example, or both. Then there would be a long series of further worked examples, and Hatch would write in the answers and sometimes the working, in ink. Occasionally he was allowed to write out the question for himself as well.

Usually young Isaac got the answers right, sometimes the teacher noticed mistakes. There may have been some rough work done on a slate, but this was no neat copy book. Once he just despairingly wrote 'wrong' in the margin. The teacher showed Hatch ingenious ways of spotting his own mistakes. If, for example, a column of figures was to be added up, Hatch would begin by adding them all, arriving at a first answer. Then he would add them all except for the first one, and arrive at a second answer. Then he would add the first number in the original column to this second answer. If the result matched his first answer, all was well. If not, something was wrong.

The learning process was laborious and repetitive. Hatch completed twenty-eight worked examples on 'The addition of money', and similar numbers for each of the five other cases of addition and subtraction on money, Troy weight, and Avoirdupois weight. The study of multiplication then began with a multiplication table—up to twelve twelves—written out by the teacher, and continued with a long series of examples of slowly increasing complexity. Hatch began by multiplying twelve-digit numbers by single digits, and ended by multiplying them by eight-digit numbers. Sometimes the page wasn't big enough for all the working.

It is a moving experience to see such evidence of an ordinary boy learning arithmetic in the reign of George I, nearly 300 years ago. Who was his teacher? We don't know—it could have been a schoolmaster, a local clergyman, or a relative. We don't know whether Hatch was learning arithmetic on his own or with others. The Hatches might have been in a position to engage a private tutor—they seem to have been owners of agricultural land, though it is not clear on what scale—but it is probably more likely that he was learning arithmetic at a school.

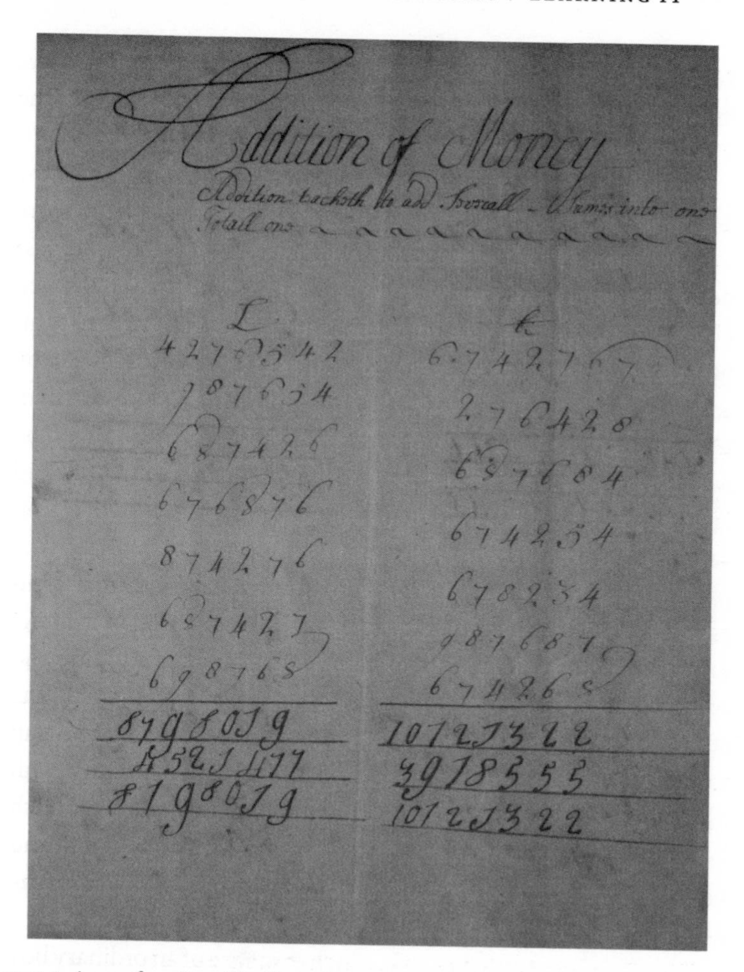

Addition of Money

Addition teacheth to add severall Summes into one Totall one

L	s
4276542	6742767
787634	276428
687426	687684
676876	674254
874276	678234
687427	987687
698765	674265
8798019	10123322
45251177	3918555
8198019	10123322

FIG 6 A page from Isaac Hatch's arithmetic exercise book, 1726
Courtesy of Benjamin Wardhaugh

Hatch comes across as having been a diligent student, but he showed no evidence of brilliance or of any particular enjoyment of arithmetic. Buried in the midst of the mostly very abstract treatment of multiplication and long division were two examples in which his teacher tried to make things more interesting for him, by linking them to what were presumably the particular interests of his farming community.

> There was a Apple wich had 7 Branches and Every Branch 9 Boughs and every bough had 34 Twiggs and every twigg had 3 Apples And every Apple was worth 3 pence. I demand the worth of such a Tree.

In similar vein, 'A Roper married his Daughter to a soper. And gave her to her portion 19 Ropes and Every Rope had 19 ends'. Hatch doesn't seem to have been enthused by this kind of thing—the attempt to liven things up did not continue.

What did he learn? By the end of the book he had been drilled to exhaustion in the four arithmetical operations. He had also learned the more difficult skills of dealing with pounds, shillings, and pence, and the systems of Troy weight (pounds, ounces, pennyweight, and grains) and Avoirdupois (tons, hundredweight, quarters, pounds, ounces, and drachms, which he called 'drames'). Each system was different, and each involved its own complications. Questions that considered the price of goods (how many pence per ounce is equivalent to a shilling per pound?) would have required some thorny conversions, but in fact Hatch's exercise book took him only as far as simple multiplication and division before breaking off.

How did this quite brief experience of learning arithmetic fit into Isaac Hatch's life? He was born in 1714 in the village of Linton, Kent. His parents were more than comfortably off: his mother made a will in the 1750s, while her husband was still alive, in which it appears that she owned in her own right some land and several houses, which were let to tenant farmers. Her husband probably owned similar property, too, and Isaac, as the eldest son, would have expected to inherit most or all of it. But the scale of the property is not clear, nor is it evident whether the Hatches themselves farmed by the time the will was made, or whether they had done so earlier in the century.

Arithmetic wasn't the first thing Isaac learned: he could already read by the time he began his exercise book, and write a little.

The very first example written out in his exercise book involved the addition of seven six-digit numbers, so presumably more basic arithmetic had already been learned, either using a slate or in another exercise book. 'Ciphering books' from this period, whether printed or handwritten, show that very basic numeracy—the reading and writing of Arabic numerals—was considered a part of literacy, and often taught alongside reading and writing.

We don't know what became of Isaac later in life. Presumably he continued to live on his family's land, and managed the work of the farmers who tenanted it. Possibly he was personally involved in that work. His arithmetic would have been put to use in his dealings with money and goods, in understanding his financial situation and the rate of production of his land. What the exercise book records is the beginning of Isaac's learning how to handle money and goods.

One charming feature which his book shared with many arithmetic textbooks of the period was the huge size of the numbers in the examples. Hatch's very first addition of money produced a result in the millions of pounds, and the apple tree in his later example turned out to have more than 6,000 apples. He was participating in a long tradition of what might be called aspirational mathematics, in which the examples conveyed, none too subtly, a message about the prosperity you could expect if you learned your sums well. In an extreme example, in 1721 one author devised a novel way to present the basics of arithmetic. *An evaluation of Solomon's wealth* presented a fairly routine course in basic mathematics, aimed at the eventual computation of the scale of King Solomon's fabulous riches.

For all that, this book did not take Isaac Hatch very far down the road to mathematical mastery. It is quite possible that later on he learned more advanced mathematical skills. Printed arithmetic textbooks enable us to see something of how his mathematical education might have continued, if indeed it did so.

In 1773 Samuel Johnson, well-known writer and compiler of the famous dictionary which appeared in 1755, made a tour of Scotland, travelling to the highlands and the Hebrides, picking up lore of all kinds and, perhaps more importantly, providing at every step material which his companion, James Boswell, would turn into his great *Life* of Johnson and his *Journal of a Tour to the Hebrides*. One night early in the tour they stayed at a house in Glenmorison called Anoch, kept by one McQueen. Johnson was struck by 'our landlord's daughter, a modest civil girl, very neatly drest'. The next morning he found himself at a loss for a parting gift for the girl, and eventually pressed upon her a book which he had picked up in Inverness and happened to have with him.

The book was *Cocker's Arithmetic*, one of the best-selling arithmetic primers of the eighteenth century. First published in the 1670s, it was in its fifty-first edition (or so) at the time, and from its pages Miss McQueen would have learned all the mathematics necessary to bring those of 'the meanest capacity' to 'the full understanding of that incomparable Art'. (Early modern title pages did not always aim to flatter the reader. Still, some capacities are meaner than others, and such language must have been demoralizing in the extreme for those who found the book hard.)

Edward Cocker himself had been a writing master, as well as a teacher of arithmetic and an engraver. In the 1750s his book was reportedly still 'much called for', although by that date it had found many an imitator, and stood out in a crowded market mainly for its longevity. Miss McQueen almost certainly received a copy of one of the editions printed in Scotland—a measure of the book's continuing popularity is that there were separate editions printed in Edinburgh and Glasgow, as well as Belfast and Dublin, during the 1750s and 1760s. The Edinburgh editor, John Mair, was a prolific author of educational works, including successful books of his own on arithmetic and bookkeeping.

Cocker's presentation was clear and logical: it began with the same mathematics as Isaac Hatch's early lessons in arithmetic. (Like Hatch's exercise book, certain of the chapters in Cocker began with a more or less impenetrable definition: 'ARITHMETIC is the art of numbering, of knowledge which teacheth to number well, *viz.* the doctrine of accounting by numbers'.) First, the 'notation of numbers', the place–value system and the names and uses of its various parts. Next, the four arithmetical operations. What followed went quickly beyond anything in Hatch's exercise book, to consider that most fruitful branch of eighteenth-century arithmetic, the 'rule of three'.

Some readers may remember the rule of three, or if not the name then the type of problem with which it was illustrated. If four yards of cloth cost twelve shillings, what will six yards cost at that rate? Cocker said the rule of three was 'not undeservedly called *the golden rule*' (the name was more usually reserved for the precept 'do as you would be done by', and there is perhaps something rather telling about Cocker's willingness to substitute a piece of arithmetic for a piece of morality), and it is easy to see why. He devoted forty pages to the basic arithmetical operations, twenty to arithmetic with fractions and seventy to operations involving or derived from the rule of three. For him, and for many other writers and their readers, the rule of three held a most indispensible place in arithmetic. Another educational writer of the eighteenth century wrote:

> By it ten thousand things are done,
> Ten thousand different ways,
> And he that learns it perfectly,
> Will merit fame and praise.

In its simplest form it dealt with situations where A is to B as C is to D, and the rule was that, given A, B, and C, you can find D by multiplying together B and C and dividing by A. Here's how Cocker stated it:

> The single rule of three direct, is, when the proportion of the first term is to the second, as the third is to the fourth...
> multiply the second number by the third, and divide the product thereof by the first...and the quotient thence arising is the fourth number in a direct proportion; and is the number sought, or answer to the question.

There were a multitude of more complicated forms of the rule for use in different situations: inverse, double, compound, the rules of 'fellowship' and of 'alligation'—some arithmeticians taught an irresistible 'Virgin's rule' of the same general type. It added up to a good deal of rote learning, but that was nothing unusual in the pedagogy of the day.

It's tempting to think that the same problems could have been solved using algebra. In some situations they might have been. But the rule of three in its many forms was part of a very long tradition of schoolroom arithmetic: it did its job well, and it also trained students to use and think about ratios. Indeed, well into the second half of the twentieth century it would thrive as a distinctive tool for solving this particular kind of problem.

Let's look at the types of problems Cocker taught his readers to solve. The example above, about yards of cloth, was one of his. For the more complex rules, and those involving fractions, Cocker ranged more widely, covering horses and soldiers that eat so many bushels of provender in so many days, cannon that use up so much powder, students who spend, and men who mow, as well as other matters of (over)consumption:

> If in a family consisting of 7 persons, there are drunk out 2 kilderkins [36 gallons] of beer in 12 days, how many kilderkins will there be drunk out in 8 days by another family consisting of 14 persons?

The recurring issue was one that we have touched on already: trade's increase, and the prodigious growth of personal wealth under the right circumstances. Usurers, investors, and merchants are all represented in Cocker's book as reaping vast harvests from the fertile fields of arithmetic: two men make a 25 per cent profit on a tun of wine; four partners build a ship; a grocer buys and sells sugar by the hundredweight. The message was plain—learn your mathematics well and you will prosper.

Here the power of the more advanced rules came into its own. The 'rule of fellowship' showed how to distribute the profits after a venture into which several partners had contributed unequal amounts, possibly for unequal periods of time. The 'rule of barter' showed how much of one product should be exchanged for so much of another. The 'rule of alligation' showed how to determine the proper price of a product formed by the mixture of others: say blended wine or mixed grains or, in one of Cocker's less modest and homely examples, gold of different degrees of purity blended so as to make a 17-carat product.

Johnson later justified his gift to Boswell. He argued that a book of science was a better companion on a journey than a book of entertainment, being 'inexhaustible'. Inexhaustible or not, that he should have thought it would yield stuff of interest and use to a country girl in the Highlands speaks volumes about the position that basic mathematical learning had attained in eighteenth-century Britain. But not everyone had Samuel Johnson to give them an arithmetic book. Where were such things learned?

In the country parish of Meppershall in Bedfordshire, one of the parishioners left money in her will to found a small school—a 'charity school', as such things were known. Her name was Sarah Emery. After the death of her sister Elizabeth, her estate was to be placed in trust 'for the schooling poor children' in Meppershall and the neighbouring village of Ampthill. In March 1692 a charity, known as the Charity of Elizabeth Emery, was created, and a school was founded at the parish church in 1698, known to some as 'Mrs Emery's School'.

This is a familiar story from this period, when such charity schools were a relatively popular form of foundation. From the records that survive, we can build up a picture of the kind of experience this institution provided. The original rules of the Emery school specified that six each of 'poor boys' and 'poor girls' should be taught, but the school quickly grew beyond that number, and over the first eight years of its existence it taught thirty-eight pupils from fourteen local families: twelve boys and twenty-six girls. Two regular 'dames' and three other occasional teachers instructed the children in their catechism and prayers and taught them to read, sew, knit, and spin. The school's purchases during those eight years include five New Testaments and no fewer than six spinning wheels. It seems to have been thriving.

The school was open during the week from Michaelmas to Midsummer, and on Sunday evenings only for the rest of the year—notice was evidently taken of the fact that many of the children would be needed for farm work during harvest time, but amendments made to the rules in 1701 suggest that truancy or lateness were nonetheless proving a problem. As well as an annual 'entertainment' for trustees and children, the accounts show an 'encouragement' given on occasion to the children both for learning their lessons and for going to church. Apparently bribery of a kind was needed to keep them in attendance.

The school was overseen by the rector of the parish church in Meppershall. Thomas Salmon was an Oxford graduate with a keen amateur interest in mathematics and music. Near the end of his life, in 1705, he would put on a musical performance for the Royal Society with specially modified experimental musical instruments, to demonstrate his theories. A large family and the responsibilities of the parish kept him in what he once ruefully called 'this little village' for most of his time, however, and he took a diligent interest in the running of the school.

While the dames taught the girls of the school to read, sew, knit, and spin, the boys were supposed to learn to 'read write and cast account'. Quite probably they learned such things from Thomas Salmon, who before his ordination had taught at a girl's school run by his mother in Hackney.

The Emery school at Meppershall is unusual in providing even this minimal evidence of mathematics teaching—albeit with precious little indication of what was taught—at a country school in early modern England. On the whole, we don't know much about the state of mathematics teaching even in the more prominent British schools in the seventeenth and eighteenth centuries. But the patchy evidence we have is dispiriting. Typical is the remark of John Newton in 1677 that 'he had never heard of any grammar school in England in which mathematics was taught' or of John Wallis that the study of mathematics was cultivated more among London professionals than at the English universities. The well-known case of Samuel Pepys illustrates that one could graduate from an English university with next to no knowledge of basic arithmetic. In connection with his work at the Navy Office, Pepys found himself obliged to engage a private tutor to teach him arithmetic, and his diary for July 1662 speaks movingly of working at his multiplication table—this at the age of twenty-nine and after schooling at St Paul's and a degree at Cambridge. By the end of his life Pepys himself would become a keen cultivator of mathematics, his devotion manifested charmingly in the incident in September 1693 when a highwayman who robbed him got away with various articles including five mathematical instruments.

Pepys was not alone in being keen on mathematics and mathematics education during the late seventeenth century, and the education reformers of the period regularly found space for mathematics in their ideal schools, finding motivation from the increasingly mathematical character of scientific and political life, and perhaps providing further evidence that mathematics education was not all that it might be. The antiquary John Aubrey listed the mathematical instruments to be owned by the ideal school in about the 1670s, in addition to 'the necessary Instruments every Scholar is to have of his owne, as Scale, Compasses, Sector, Protractor, & Globes'. The list ended 'Let 'em be furnished with Microscopes, Telescopes, Dark-roomes for taking ones picture or Landscapes: it will sett them *agog* (as they say)'. This was no idle fancy: in 1673 the Royal Mathematical School was set up at Christ's Hospital in London with the intention of training boys for naval careers, with contributions from Pepys, Newton, Edmund Halley, and others. Yet by the 1690s the parlous state of the Royal Mathematical School itself was a matter of remark—a correspondent of Pepys referred to sailors who didn't know the 'usefulness or sweetness' of maths, hinting at a resistance to book learning among those of the naval profession.

But the teaching of arithmetic in schools was by no means confined to specialist institutions like the Royal Mathematical School. The younger Eton boys (when they weren't poking holes in their hats to make pinhole cameras) are said to have had, by the mid-eighteenth century, two hours of writing and arithmetic even on holidays, while the older boys learned geography and some algebra. Even the pages at George II's court in Hanover were required—in theory—to learn basic literacy and numeracy as well as other skills.

For those of more modest means, a large number of individuals also ran private mathematical schools, at which a very wide range of skills could be learned:

> Writing and Arithmetick, Taught by Mr. *Joseph Pepper*, at *Stamford* in *Lincolnshire*; with whom Youth may be commendably boarded for their speedier Improvement.

This is typical, and appeared in Francis Perkins' *New Almanack* in 1722.

There seems to have been sufficient demand for mathematically trained young men, and therefore for mathematical training for young men, to sustain a large number of such mathematical schools, both boarding and day. To that number may also be added the so-called 'dissenting academies': schools run by the nonconformist communities, which flourished from the second half of the seventeenth century onwards and which tended on the whole to teach more rather than less mathematics, concerned as their founders often were with the reform of practical life and the inculcation of virtuous—and incidentally profitable—diligence in their charges.

The strength of nonconformity's commitment to the teaching of mathematics is vividly illustrated by an incident from 1714. The Schism Bill, which aimed to prevent teaching by nonconformists, was amended in the Lords due to pressure from mathematical teachers and practitioners. The amendment specifically exempted from the ban on teaching

> any Person, who as Tutor or Schoolmaster, shall instruct Youth in Reading, Writing, Arithmetick, or any such Mathematical Learning only, so far as such Mathematical Learning relates to Navigation; or any Mechanical Art only.

And for those more humble still there were the schools—more and more of them—set up as charitable foundations, like the Emery school in Meppershall. About their mathematical teaching—if any—we know very little indeed. The pattern at the Emery school in which (at least in principle) boys were taught some arithmetic but girls were not, seems to have been common in theory. Yet—as we will see in more detail in the next chapter—a moment's reflection suggests that if numerate skills were needed, they were needed by women just as frequently as by men.

The pioneering statistician William Petty had suggested in 1685 that 'one day Arithmetick and Accountantship will adorn a young woman better than a suit of ribbands'. Later in the eighteenth century Mrs Primrose in Goldsmith's *Vicar of Wakefield* would boast that 'my two girls have had a pretty good education'. In addition to understanding 'their needle, bread-stitch, cross and change, and all manner of plain-work' they could 'read, write, and cast accompts' (like the boys at the Emery school, but not like the girls). Even Mrs Malaprop's list of what she would like her daughter to learn included 'a supercilious knowledge in accounts', although she remained suspicious of 'algebra, or simony, or fluxions, or paradoxes, or such inflammatory branches of learning' and of 'your mathematical, astronomical, diabolical instruments'.

The skills of domestic economy and management which were expected of many women were increasingly numerate ones, and despite the persistence of some ugly prejudices ('oft we hear, in height of stupid pride, / Some senseless idiot curse a lettered bride', wrote Elizabeth Tollet in 1724 in her *Hypatia*), practical necessity must have resulted in a great many women possessing basic arithmetic who possessed little else from the world of the sciences.

Yet many were the schools of the eighteenth century where, as at Meppershall, the rules made no provision for teaching arithmetic to girls at all. One example was the 'Protestant Dissenters' Charity School' in Spitalfields. Founded in 1717, by the late eighteenth century it was teaching forty boys and thirty girls from ages eight to eleven. 'The Boys are taught Reading, Writing, and Arithmetic; the Girls are taught to Read, Write, Knit, and Sew'. All were also taught the Christian religion. Similar examples could be multiplied, particularly in London, where charity schools were both numerous and relatively well documented. A provincial example was the charity school of Baldwin Street in Bristol, founded in about the 1750s, whose purpose was avowedly to teach poor girls 'to read and spin' and which made no provision for their numerate education. Yet at the same time the rules of that school did expect its girls to participate in—and presumably have some understanding of—a complex system of payments and bonuses for the spinning they did there. Did this amount to a minimal sort of arithmetical training in itself? One possible explanation is that teaching 'to read' routinely included teaching children to read and handle numbers: that basic numeracy was treated as part of literacy. Another is that the convention that girls did not learn arithmetic was honoured more frequently on paper than in reality. Unfortunately we do not know for sure.

An exception from late in the eighteenth century, when things were perhaps changing in this respect, was the girls' charity school of St Pancras in London. The purpose of the school, instituted in 1776, was to prepare 'the female children of the industrious poor' for domestic service, training them from the ages of eight to thirteen. A set of rules published in 1796, a time when the school was thriving, specified that the girls 'after the Age of Eleven Years, be taught to Write a legible Hand, and the two first Rules of Arithmetic'. This was in addition to a large quantity of other work, with the emphasis on plain needlework, reading, and the principles and duties of Christianity. The 'writing master' attended for two hours twice a week to teach writing and arithmetic, and the girls' writing and 'Cyphering' was to be inspected by the school's board monthly.

For boys, who apparently had more access to numeracy much earlier, there survives in some cases detail about what arithmetic or mathematics they were taught. The charity school for boys set up at York in 1705 intended its pupils to go on at the age of twelve to be apprentices 'to sea, or some Mechanicall Trade, if not fit for Sea', and for this they were judged to need 'the first Rudiments of Arithmetick' in addition to reading, writing, and the catechism of the Church of England. The master was explicitly required by the school's rules to be competent in this respect: he must understand 'the Grounds of Arithmetick' and teach the boys 'the first five Rules, to fit them for Services or Apprenticeships'. Those were presumably the same five rules that Isaac Hatch learned: numeration, addition, subtraction, multiplication, and division.

Here mathematics was, apparently, at par with the other studies in at least some respects. One rule specified that:

> The Master shall give the Boys some Exercise every Saturday-night, or Eve of an Holyday, either to get part of a Psalm, or practicall Chapter by heart, or to write it down; or else a Question in Arithmetick, such as may employ their Minds at vacant hours.

What kind of arithmetical problem would have required effort comparable to the memorizing of a psalm or a prose passage? By the end of the century the school had a headmaster, Thomas Crosby, who was himself a mathematical practitioner. He edited an arithmetic textbook which contained a selection of suitable problems:

> In 7 cheeses, each weighing 1 cwt. 2 qrs. 5 lb.—how many allowances for seamen may be cut, each weighing 5 oz. 7 dr.?

> There are three numbers 17, 19, and 48—I demand the difference between the sum of the squares of the first and last, and the cube of the middlemost.

The book was the wildly successful *Tutor's Assistant* by Francis Walkingame. Devised by the original author for his own school in Kensington, it covered arithmetic, fractions, decimals, and duodecimals, and featured collections of questions on each topic. It was first published in London in 1751 when Walkingame was in his late twenties, and reprinted nearly 300 times up to his death in 1783. There were editions in several other English cities, including York from 1797, and also in Canada. Together with Thomas Dilworth's comparably successful *Schoolmasters Assistant: Being a Compendium of Arithmetic* of 1744 it was one of the two dominant arithmetic primers of the period. As late as the mid-nineteenth century it was still said to be the most used of all school arithmetic books.

Writes Ruth Wallis, historian of eighteenth-century mathematics education:

> He [Walkingame] aimed to equip his pupils to solve problems by using formulae, without necessarily understanding their validity. His success was based on inclusion of contemporary commercial topics, worked examples, and many practice questions with answers.

The final selection of questions his book contains gives some idea of the kind of material that might have been handed to the York pupils—and many like them—on Saturday nights. Most merely required calculation:

> What is the value of 14 barrels of soap, at 4 1/2 *d. per lb.* each barrel containing 254 *lb.*

> What number added to the cube of 21, will make the sum equal to 113 times 147?

But the book also contained material, on compound interest and other matters, in which basic algebraic notation was introduced and used:

> The letters made use of in Compound Interest are,
> A. the amount.
> P. the principal.
> T. the time.
> R. the amount of 1*l.* for a year, at any given rate ...
> When P, T, R, are given to find A.
> Rule. $p \times r^t = \mathrm{A}$.

The preface to the book gives a rare glimpse of schoolroom practice in this period. Walkingame compiled his collection of questions in order to save himself the labour of 'writing out the Rules and Questions in the Children's Books' either during the lessons or between them. With a manuscript book of questions written out, he could simply hand it to whichever student was in need of material, and require the student to make a copy of the necessary section. But of course, having just one book resulted in a queue, and Walkingame's solution was to have his book printed.

He suggested that it would thus be of use not only to his own school in Kensington but to other schoolteachers 'as would have their Scholars to make a quick Progress':

> It will also be of great Use to such Gentlemen as have acquired some Knowledge of Numbers at School, to make them the more perfect;

> likewise to such as have completed themselves therein, it will prove . . . a most agreeable and entertaining Exercise-Book.

Apparently, at least some of those groups agreed with Walkingame in their assessment of his book over the next hundred and thirty years.

Mathematics was not only learned by young children. For some professions—as for that of Samuel Pepys—it was, or ought to have been, a part of professional training.

One individual who had such an experience was Robert Gardner. His handwritten 'Book of Acompts', made in Lancashire in the 1770s and 1780s, began with some quite sophisticated mathematics. In other respects he was a little like Isaac Hatch: he was reasonably careful in his presentation of 'rules' for performing various operations, but rather less so about how he set out his worked examples.

The contents of his book began with 'alligation medial', a rule to use when different substances were to be mixed and a fair price found for the mixture. For example, 'Suppose 15 Bushels of Wheat at 5s the Bushel & 12 of rye at 3s 6d the Bushel were mixt togather what is the mean Rate or price a Bushel may be sold for without Gain or Loss[?]' Once again, no algebra was employed, just a verbal rule: 'As the sum of all the Quantities is to the sum of all the Rates so is any part of the mixture to the mean rate or price of that part'.

This was quite advanced stuff (this was not a 'book of accounts' or of accounting, whatever the title page may have said), and it is evident that this was not Gardner's first experience of mathematics: he might have first filled one or two exercise books, or more, with easier material. The contents continued through the rule of 'false position' (which found an unknown value by taking a guess at it and then correcting the guess using what amounted to the rule of three) and the extraction of square roots, before moving to the fields of measuring and geometry, in what seems to have been the normal progression for those who got so far in their study of mathematics. The subjects covered were very practical ones: 'To measure windows' and 'To measure Pavements', 'Land Measure', and the measuring of irregular shapes composed of multiple triangles. It seems that Robert was being groomed for the work of a surveyor or perhaps that of a more general mathematical practitioner. The training was attentive to distinctly local matters, since some questions were to be solved 'in Lancashire measure' and others in 'Westmoreland Measure'. Sadly the difference between the two was not made at all clear.

Like Hatch, Gardner may have been learning at a school or in private—the sometimes scrappy state of his work does not suggest close oversight by a tutor, but it's impossible to be sure. According to an inscription in the middle of the book he lived at Cockerham in Lancashire in 1787, when some of the exercises were carried out. His was a very common name in eighteenth-century Lancashire, and we cannot be certain of his story. But if he was the Robert Gardner who was baptized in Cockerham in August 1765, he was 21 or 22 when most of these exercises were done. As was the way of these things, the book also seems to have been used by his brother John, and a 'Richard' also signed his name on the front cover, perhaps claiming ownership—or trying to—at a later date.

Isaac Hatch learned mathematics as a child, while Robert Gardner seems to have been studying it in his early twenties. But some writers suggested that their textbooks might be used by older students, too. What might an adult do, who was interested in beginning or improving mathematics during this period?

One answer was to join a mathematical club. Spitalfields Mathematical Society, for example, existed from 1717 until 1846, and for much of that time it facilitated mathematical learning by working men in east London through a combination of lectures, problem solving, and mutual instruction. It wasn't unique: we know of mathematical societies in Manchester (founded 1718) and Northampton (founded 1721), and we have very scanty references to at least three more in London in the first quarter of the eighteenth century, and to others in Lewes, Wapping, and perhaps York. Clubs of all kinds were an extremely popular type of activity, and one contemporary estimated that they involved up to 20,000 men every night in London alone.

The Spitalfields society was founded by Joseph Middleton, teacher of mathematics, marine surveyor, and author of a 'Course of Mathematics' which circulated as a manuscript. It met at first in a series of pubs: the Monmouth's Head, the White Horse, and the Ben Johnson's Head, all in Spitalfields. The Society's later habit was to meet for three hours on a Saturday evening, which may have been the pattern from the beginning. Members paid to attend, and by the 1780s it was fourpence a meeting.

Attendance was in the fifties and sixties at first, and the members were tradesmen: weavers, brewers, braziers, bakers, bricklayers. Most were local men, but some apparently travelled from elsewhere in London to attend the meetings once the Society's reputation had begun to grow.

We don't know quite what form these early meetings took, but a reference to the members' 'slates' suggests that for part of the meeting Middleton was teaching his course of mathematics to the other members. When a constitution was drawn up in the 1780s, it specified that for one of the meeting's three hours silence was kept,

> and every member present shall employ himself in some Mathematical exercise...And if any member be asked a question in the Mathematics by another, he shall instruct him in the plainest and easiest method he can.

There were penalties for failure in this regard, as there were for breaking silence and such other offences as 'behaving riotously, or using abusive language'. Although smoking and drinking were permitted (as the nineteenth-century mathematician Augustus De Morgan would much later put it, 'each man had his pipe, his pot and his problem'), it was forbidden to 'curse, swear, game, or lay a wager'.

We don't know what the mathematical exercises were that the men of the Spitalfields Society worked on during their hour. Beginners might have been given questions like those in Walkingame's textbook, while an idea of what might have been done by more advanced members can be gained from a publication called *The Mathematician*. This magazine ran from 1746 to 1751 and was associated with a 'Society of Gentlemen' which apparently met every Saturday evening at the home of a Mr John Pelah 'in Bunhill Fields near Moorgate, London'. It consisted of problems and solutions, and a preface promised that 'no Problems [will] be proposed here, but such as are possible, rational, and, as far as may be, applicable to some Use or Purpose in Life'. Questions in the first number included these:

> If a Cubic Foot of dry *English* Oak be put into a sufficient Quantity of fresh Water, how much of it will be immerged, and how much emerge, and what Weight must be laid on it, to make it level with the Water's Surface?
>
> How high must an Eye be elevated above the Earth's Surface, to see two fifth Parts thereof; allowing the Earth to be spherical, and its Circumference 25,020 Miles?
>
> To draw a right Line through a given Point, terminating in two right Lines given by Position, so as to be the shortest possible.
>
> At a certain Place in North Latitude, the Sun was observed to rise exactly at 3H. 58 Min. and at Six o' Clock his Altitude was taken the same Morning, and found to be 15°: 20′, his Declination being then North; required, the Latitude where, and Day of the Year when, these Observations were made.

This was demanding stuff, and the magazine also carried serials setting out the history of geometry and the theory of conic sections. To judge by the recurrence of a small number of names in the credits for solutions, *The Mathematician* did not attract a very large number of readers, although the editors attempted to encourage contributions even from the more timorous by a promise not to print incorrect solutions. Presumably the society whose work it recorded remained small—we do not know what happened to it after the magazine ceased publication in 1751.

For the working men of the Spitalfields society the study of mathematics was a form of self-improvement whose results could be quite spectacular. Thomas Simpson, for example, who was working in Spitalfields as a weaver in the 1730s, would become a mathematical teacher and author following intensive contact with the Society. Eventually he became a Professor at the Royal Military Academy at Woolwich, one of the most prestigious appointments in practical mathematics in Britain (we'll hear more about Woolwich and its professors in Chapter 7). John Canton followed a similar path from weaver to teacher: he remained in Spitalfields, where he ran his own school.

In the longer term the Spitalfields Mathematical Society moved away from its roots both socially and in terms of subject, so that by the early nineteenth century it was struggling either to recruit members from the class for which it was founded or to compete with other bodies such as the London Mechanics' Institute, founded in 1826. During the 1770s and 1780s it absorbed at least two other societies, with its interests broadening into the study of science and history as a consequence. Members now gave lectures on a rotating schedule, and on Newton's birthday medals were awarded for the best. There were repeated attempts to open some of the lectures to the public, which attracted audiences of up to 300 on occasion and seem often to have made the Society a profit (despite a threat of legal action at one stage for unlicensed entertainment). The Society's credit and reputation were apparently such that it could on occasion certify to a prospective employer that a given member was suitably qualified in mathematics.

Increasing respectability saw the Society move into new premises: what had once been a Huguenot chapel. But by the early nineteenth century the period of silent work at problems seems to have been dropped, and the offences for which fines could be levied now included introducing into a lecture 'any controversial points of divinity or politics'. Evidently the tone had changed quite considerably, and the social make-up shifted away from the trades and towards professional men: surgeons and lawyers, as well as masons and dyers.

Despite a library of 4,000 volumes, a large collection of instruments—the collection ranged from a slide rule to telescopes, air pumps, globes, and even a Chinese abacus—and a tone which, according to a contemporary, 'cannot offend the most punctilious', the Society could no longer attract members or lecturers in sufficient numbers, and in 1846 it finally merged with the Royal Astronomical Society, its sixteen remaining members (no longer working men but barristers, stockbrokers, and surgeons) being granted life memberships of that Society.

Around the time the Spitalfields Mathematical Society was entering its final decline, another student was learning mathematics. Her name was Ann Mohun, and her story brings together much of what we have seen in this chapter.

Her mathematical education took place at the school in Ingleby in North Yorkshire ('one of the wildest parts of the district of Cleveland' according to a nineteenth-century topographer). We don't know for certain who her parents were or what they did, although it's likely they were farmers or farm labourers—the sort of folk whom Isaac Hatch's parents had as tenants on their farms. It seems they may have moved to the village after Ann was born.

In some ways her mathematical exercise book tells a story similar to that of Isaac Hatch eighty years earlier. The strategies passed down through generations of teachers have a logic of their own, independent of growth and change in research mathematics. Like him and many others, she used her mathematics workbook for pen practice almost as much as for arithmetic: she wrote her name repeatedly and wrote out the date on several different pages, recording for posterity the fact that she used this book during the year following February 1808. She was probably between about eight and fourteen, like the students in the schools discussed earlier, although perhaps on the older side judging by the quality of her handwriting. She filled spare space on the page with ink doodles and sketches of leaves, and her unruly use of the paper suggests that her work was not closely supervised.

But there were other times when she took care over the presentation of her work. She provided the section headings with coloured backgrounds, using watercolours: yellow, mostly, but there is also some blue on certain pages. And she used red ink to underline important results and to set off parts of the text from one another.

What did she learn? The book didn't begin at the beginning in the way that Hatch's did: it began with a table showing how to convert quantities of wool and hay from one system of weights to another. It continued somewhat haphazardly through arithmetical operations and further discussion of weights and measures. There were no opaque 'definitions', but operations were often introduced with a 'rule' or a description of how to carry them out. It seems that Ann already knew how to count and add up when the book was started—slightly more than Isaac Hatch, in other words—but she had yet to learn how to perform long subtractions and multiplications.

As with the other examples we have seen, there was an emphasis on manipulating large quantities of money and other goods, and on the 'reduction' of quantities given in one form to another: how many farthings are there in twenty-one guineas, for instance. There were long sequences of examples of these operations, as well as such specialized mercantile matters as tare, tret, and cloff:

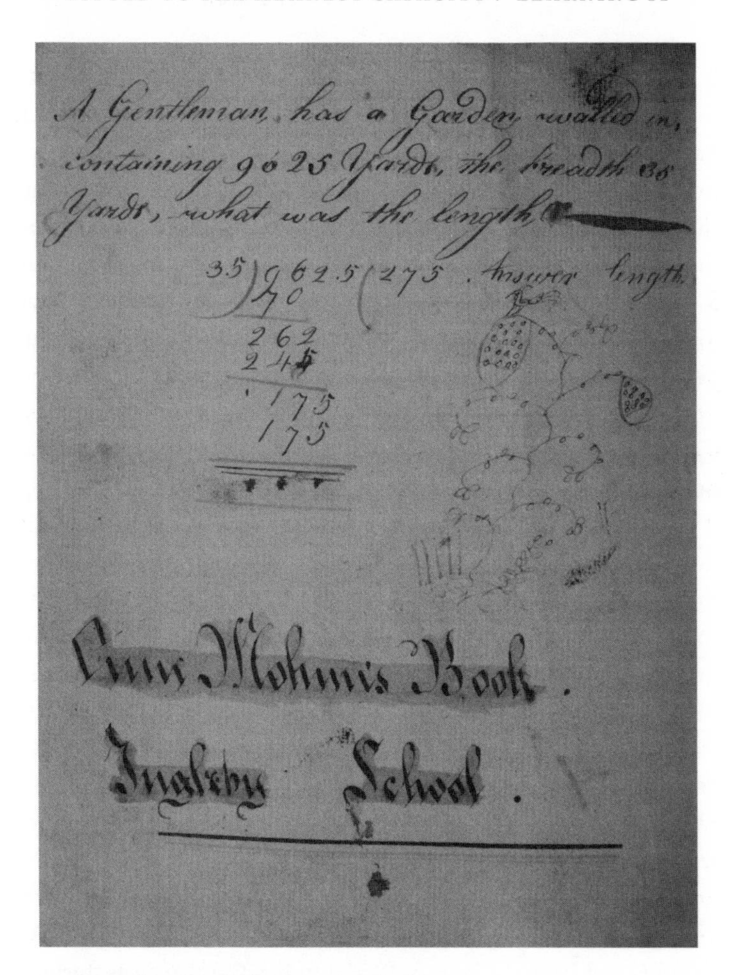
A Gentleman has a Garden walled in, containing 9625 Yards, the breadth 35 Yards, what was the length

35) 9625 (275 Answer length
70
262
245
175
175

Ann Mohun's Book.
Ingleby School.

FIG 7 A page from Ann Mohun's exercise book, 1808
Courtesy of Benjamin Wardhaugh

> Tare is an allowance made to the buyer of any commodity for the Weight of the Box, cask, &c ... Tret is an allowance ... on account of waste ... Cloff is an allowance ... to make the Weight hold out when retailed.

There were also some interesting examples of calculation involving dates, sometimes quite surprising ones:

> America was discover'd in the Year of our Lord, one thousand two hundred and eighty one, how many Years since?

Ann's teacher was apparently no historian.

Like Isaac Hatch, Ann was taught to check or 'prove' her calculations, using a variety of strategies. A simple subtraction was 'proved' by adding the result to the number subtracted: if the result matched that in the first line of the calculation, all was well. Ann had a penchant for writing out the answer on a jubilant diagonal in that case, producing a most striking effect on the page.

The last things in the book were exercises on 'Ale and Beer measure' and the measure of wines and spirits. Perhaps the mathematical instruction was in fact conceived with the boys in the class mostly in mind—it certainly retained its aspirational focus to the last:

> A gentleman ordered his Butler to Bottle off ⅔ of a Pipe of French wine into quarters and the rest into Pints I desire to know how many Dozen of each he had.

A pipe of wine is 126 gallons.

Ann's was a chaotic and complex book, and it bears witness to a process of mathematical learning much more sociable than that of Isaac Hatch—it sometimes recalls the milieu of mutual instruction and self-help cultivated by the Spitalfields Mathematical Society in its early years. Ann herself wrote out problems in a dark ink, and their solutions in a light one. There was rough work in pencil, some headings were in a larger form of Ann's ordinary hand, and some in a special copperplate-style script. Over the months Ann's own handwriting slowly improved, and there were times towards the end of the volume when she achieved a really admirable appearance on the page.

Yet she had classmates, or perhaps siblings, who had less respect for the written page. John Bentley obtruded his name at more than one point for no obvious reason—was he helping Ann with her schoolwork and then signing his name at the foot of the relevant page? 'William Mo' did the same: perhaps William Mohun. At one point there broke in a more childish hand than Ann's, writing in thick letters at the foot of the page:

> And if you dont do as I tell you I send you to sea and that will make you A Good boy.

And on another occasion Ann herself wrote out the following splendid instruction:

> Run and Jump and talk aloud in the fields but in the house and among your parents and friends Be quiet and Sedate [sit down] William

Once or twice pages were cut out of the book, perhaps where the chaos on the page had got out of hand, or where either Ann or an adult wished to remove the evidence of error or inattention in class. One indiscretion which escaped the pruning knife was a passage of text that began:

> If your Heart be your own, I demand it as A new year's Gift... and be you sure that I have nothing, I say nothing which I ought to refuse as a Recompence for a Present which whill be so Dear to me.

Sadly, we have no record of the generous offer's result.

More enigmatic still, the final page is signed 'Ann Elcoat's book', in a more grown-up handwriting. That same hand added a few extra examples and corrections elsewhere in the book. In this same hand there is, inside the back cover, a date in 1844 and a list of place names in the North of England. What had happened?

An Ann Mohan married a Luke Elcoat in Durham in 1829. Assuming this was our Ann Mohun, we should perhaps imagine her—older, married—going back over her mathematical exercises and correcting her young self. (She had new interests by this time—a note about the 'German polka' seems to date from this married period.) The book was still 'Ann Elcoat's Book' after a further move to Stockton-on-Tees, but it would eventually become 'Nancy Elcoat['s] Book' in a new handwriting. Presumably Nancy was Ann's daughter, and the book had been handed on.

This exercise book tells a remarkable story, of a young Yorkshire girl born in the last years of the eighteenth century who was afforded the—by no means universal—privilege of attending the mathematics classes at the local school and who struggled against distractions of various kinds to take down and complete the exercises that were asked of her. She received a basic arithmetical education of the kind that had remained substantially unchanged for a century, and that would have been thought fitting for a country boy or (possibly) girl at many a time and place in eighteenth-century England.

Some days she decorated her book, 'illuminating' the mathematical notation with watercolours and setting out titles and divisions neatly. On other days she despaired of neatness and drew pictures of plants in the corners. And on other days again she resorted to reprimanding her classmates in writing, presumably because she couldn't talk in class without getting into trouble herself.

No-one seems to have been taking much notice of what she wrote here—at least, it's hard to imagine a teacher letting her get away with an exercise book of this appearance had it ever been inspected. Was that because, since she was a girl, her mathematical education was not attended to seriously? We cannot know.

More remarkably still, she kept the book with her, and after her marriage twenty years later she wrote her name in it again, looked over the mathematical exercises and in places added to them with new material. Apparently what she had learned in the schoolroom in Ingleby was still useful to her as a married woman in Durham: perhaps less so the conversion of hogsheads of ale into pints than the rules for setting out monetary calculations and accounts. And the book eventually passed to her daughter, who was interested enough in the material it contained to insert her name at two points and make some other small additions of her own.

Like all the material in this chapter, this is a story of mathematics learned for a reason, and of simple arithmetical strategies cherished for their power to change and improve one's life by being useful. But what was mathematics useful for, to someone like Ann Mohun?

Chapter 4

'My Scarbrough expenses'

USING IT

My Scarbrough expenses			
gaue to the porre people at york	0	1	0
gaue at molton	0	0	6
gaue to the Charrety Childer & at the Cong. Room	0	9	6
pade at the long Room	0	5	0
for the hot Baths	0	10	0
gaue to the saruant	0	1	0
for going in to the sea	0	3	0
...			
for a pint of wine	0	1	0
...			
for a Fann	0	1	0
for a plaything	0	4	0
8 pound of Coffe	1	1	0
...			
march the 25 1725	23	0	2

This account was made by Margaret Frank, in her teens but apparently mistress of her household following the death of her mother and the marriage of both her sisters. She noted down carefully what she spent and where, and also what she received in rent from tenants and other sources. For the purpose she used a book that already had a long and fascinating history of its own.

The first user of the accounts book had been Margaret's great-grandmother, Elizabeth Hare or (during her first marriage) Elizabeth Leigh. Living in Surrey in the 1630s and 1640s she kept a regular account, in her words a 'Knote of what money I haue disbursed for my husbans ocations or ealse deluered back too him by his apointment'.

She listed disbursements, for example, 'for buying of dogs-meat' (£1), for 'bran for the dogs' (10s 4d), or 'for a mans work six days about the dog kenell' (6s). She also reimbursed her husband for small expenses he had incurred (many of them connected with 'the dogs'): it seems that she was in charge of the household's cash. As well as on the kennels, money was spent on silverware and fabric ('too ells of fine holand for a halfe shirt', 'three peeces of duck fustian towards a bed'), on tailors and barbers, and of course on food ('halfe a buck', 'chickins', 'more chickins'). A typical year would show perhaps twenty different payments in all, but the pattern varied widely. In some years there were such sequences of expenses as 'my midwife', 'for churching mee', 'the babies feast', and 'a cake when I lay in'. The individual payments were not dated, but at the end of the year, or sometimes the quarter, they were added up to show 'the sum totall of what I haue layed out for my husbands owen ocations... or returned too him by his apointment or Laied out about houskeping'.

After the death of her first husband, Elizabeth returned to her parents' home in Norfolk for a time, where she does not seem to have used the accounts book. After her second marriage, to Sir John Lowther of Lowther, first baronet, she took up detailed household management again in her new home in Westmorland. She now had a somewhat larger establishment and added up her totals quarterly, but otherwise she continued using much the same simple system as before. She went on doing so until the 1670s, probably until her husband's death in 1675.

After that she lived with her eldest son, Ralph Lowther (there were older Lowther children from Sir John's previous marriage, so Elizabeth's children held no titles). During the later 1670s and the 1680s the book was in use for the calculation of rents received and due, and for compiling inventories of household goods, clothes, and tableware: '1 nue sute of eged night close with singel rufels to them 6 pare of durty Gloues of Pegey'; '2 bibles 7 phamflitts 12 books beside'. During this period the book's appearance became quite chaotic, and the chronological sequence is not at all clear. Presumably it was clear to Elizabeth, even if the notes she was making were no more than aide-mémoires: they no longer seem to have had the function of an orderly account book.

Elizabeth died at Ackworth Park in Yorkshire in June 1699, aged 79. By this time her handwriting had changed almost out of recognition, and it's hard to be sure exactly which parts of the book were written by her. It was also used in the 1650s by one of her stepsons for a brief inventory of clothes, and by one of her stepdaughters to record recipes. At least two other hands later continued the collection of recipes and medical instructions, including that of Elizabeth's granddaughter Elizabeth Frank.

This younger Elizabeth also briefly used the book to keep accounts, during 1713. Her responsibilities seem to have been more limited than those of her grandmother, and she listed mainly small sums spent on items of clothing, fabric, and mending for herself and her three children.

One of those children was Margaret, whose trip to Scarborough began this chapter. The accounts book now recorded the domestic arithmetic of three generations of women, and was in a very messy state when she received it. She used the remaining blank and partially blank pages for her own accounts. They began in about 1728, when she was probably in her teens, and continued through her marriage in 1738 to Richard Frank of Campsall Hall in Yorkshire.

She kept tradesmen's receipts—for example for shoes bought by Margaret and her sister—lists of clothing, and what appear to be lists of drinks supplied to guests ('new goosberry Wine 40 pints'). She also wrote out copies of receipts in her own hand. After the death of her mother in 1726, and the marriage of her two older sisters, it appears that Margaret was in charge of at least part of the household's money despite her young age (her exact date of birth is uncertain). The trip to Scarborough, and an earlier one to York, saw her recording the sums she had given to 'the Charety Childer' and 'the porre people' as well as paying a fee 'for going in to the sea' and 'for Chare hire'. Ribbons, handkerchiefs, girdles, wine by the pint, and coffee by the pound: the impression of frivolous spending continuing into early adulthood is offset by the careful recording of monies (sometimes identified as rents) received:

> Receiued Aug. 3. 1729 of richard Tomkins the sum of seuen pound ten shillings in full for a Quarter Rent for his House due at midsummer last past I say receued per me M.F.
> Receued Aug 1 of Mr John Swift six pounds four shillings in money & goods which is the full of all accompts whatsoeuer from the said john Swift to this day I say Receued in full per me

This remarkable volume, now in the archives at Doncaster, bears witness to several generations of domestic arithmetic by the women of a modestly prominent family. By Margaret's time it had become a complex, multi-layered record more than a century old.

The book illustrates one of the most pervasive uses of numerate skills in early modern Britain: to record and analyse monetary transactions. The life of almost anyone in a money economy must involve some exercise of numerate skills, if only to add and subtract, to recognize that fair sums have been paid and correct change given. Dealing with wages, rents, or other regular payments—from either point of view—might involve the occasional multiplication or division. As the quotations from Elizabeth Lowther and her great grand-daughter show, only minimal literacy need be involved in writing such things down: neither woman spelled in such a way as to suggest much contact with printed text. Many who were numerate may not have been literate at all, or have kept any written record of their transactions (tally sticks were still a current technology at the Exchequer, after all, and would remain so until 1826). It is more common for the evidence of financial record-keeping to be much more scanty than Margaret Frank's book, consisting of jottings in the middle of a diary or on the flyleaf of a printed book.

The financial responsibilities and record keeping of Elizabeth Lowther or Margaret Frank may have been unusual, at least in the density of their written record, but those two women were by no means alone in doing such things. If Elizabeth Thomas, writing *The True Effigies of a Certain Squire* in 1722, could still have the squire remark that there was no need to learn arithmetic ('I know by custom two and two is four; / My man is honest, then what need I more?'), for those who did not have a 'man' to do it for them there *was* a need, and a need that often fell to the women of the household.

Thus, by the mid-eighteenth century printed manuals of basic accounting were certainly reaching a female audience. The 1755 *Plain and Easy Treatise of Book-keeping* by Richard Shepherd, for instance, had thirty women among its subscribers. Hannah Glasse, better known for *The Art of Cookery, Made Plain and Easy* (1746) and for her costumier's shop on Tavistock Street in London, also authored an account of domestic arithmetic aimed at women. After setting out under the title of the *Servant's Directory* 'the Duties of the Chamber-Maid, Nursery-Maid, House-Maid, Landery-Maid, Scullion, or Under-Cook', her book moved on to address itself 'to the Young Housekeeper', for whom were set out 'Directions for keeping Accounts with Tradesmen'. She gave detailed explanations of notation ('s stands for stone, p for pound, and q for quarters') and of a proper way to lay out accounts on the page, recording sums spent and prices per pound of the goods bought:

> At any time you may look over your Account here, and in a few Minutes know to a Farthing what Expense you are at for House-keeping, and everything else; then mind to carry your Sums forward to the next Month, that you may know, at the Year's end, what you have paid to every Tradesman.

Next she set out blank accounts, ready to use, for a year's payments to butcher, baker, poulterer, fishmonger, cheesemonger, and 'Butter-man', and so on down to the chimneysweep's bills. Separate forms were provided for the recording of such 'Ladies necessaries' as the stay-maker and the hair-cutter, with spaces to record 'Where the Trades-people live'. Together with a simplified set of blanks for smaller households, the whole collection of blank accounts occupied nearly 400 pages.

It was a much more elaborate scheme than that employed by Margaret Frank, but it was an indication of the growing importance of numerate skills in everyday life, and the growing sophistication with which they might be used. (Glasse herself, rather ironically, was imprisoned for debt for part of 1757—the penalties for failure in domestic accounting could be serious.) Glasse reflected that trend by finishing her volume with a series of ready reckoners for wages and simple interest, tables of conversion for weights and measures, a table of the rates charged for various forms of transport, and even a tide table.

At the same time that Margaret Frank was keeping her accounts in the 1720s and 1730s, one John Smart ('Gent') published a set of *Tables of Interest, Discount, Annuities, &c.* in 1726. They show us what those who did it for a living thought about account keeping.

A copy of the book survives which was given by the author to one Henry Metcalfe, who later passed it on to his older brother Thomas (possibly an ancestor of the baronets Metcalfe). In 1754 a new owner noted on the flyleaf that it had been given to him 'some time ago'. That new owner was the twenty-two-year-old Frederick

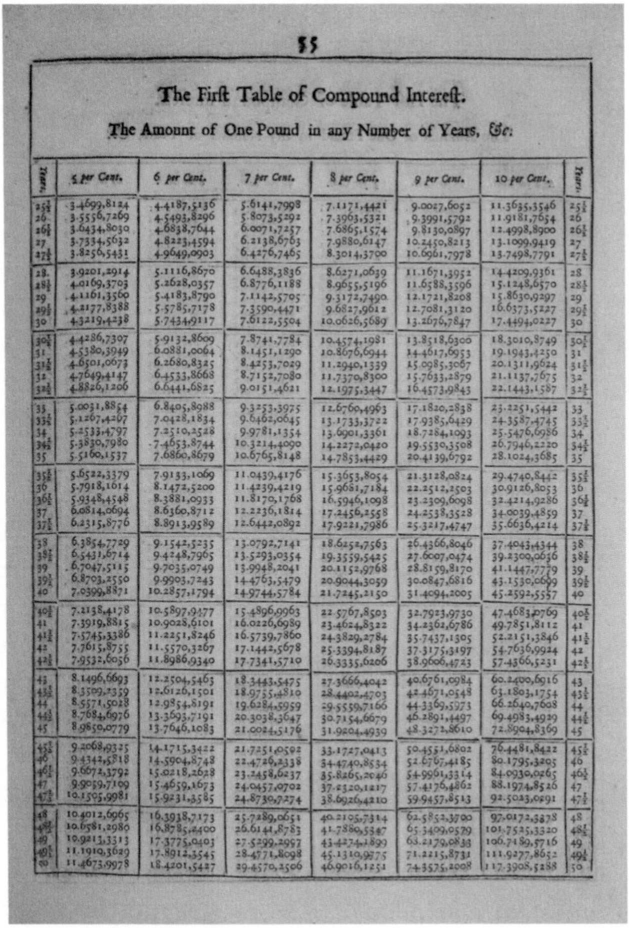

55

The First Table of Compound Interest.

The Amount of One Pound in any Number of Years, &c.

Years.	5 per Cent.	6 per Cent.	7 per Cent.	8 per Cent.	9 per Cent.	10 per Cent.	Years.
25½	3.4699,8124	4.4187,5136	5.6141,7998	7.1171,4421	9.0027,6052	11.3635,3546	25½
26	3.5556,7269	4.5493,8296	5.8073,5292	7.3963,5321	9.3991,5792	11.9181,7654	26
26½	3.6434,8030	4.6838,7644	6.0071,7257	7.6865,1574	9.8130,0897	12.4998,8900	26½
27	3.7334,5632	4.8223,4594	6.2138,6763	7.9880,6147	10.2450,8213	13.1099,9419	27
27½	3.8256,5431	4.9649,0903	6.4276,7465	8.3014,3700	10.6961,7978	13.7498,7791	27½
28	3.9201,2914	5.1116,8670	6.6488,3836	8.6271,0639	11.1671,3952	14.4209,9361	28
28½	4.0169,3703	5.2628,0357	6.8776,1188	8.9655,5196	11.6588,3596	15.1248,6570	28½
29	4.1161,3560	5.4183,8790	7.1142,5705	9.3172,7490	12.1721,8208	15.8630,9297	29
29½	4.2177,8388	5.5785,7178	7.3590,4471	9.6827,9612	12.7081,3120	16.6373,5227	29½
30	4.3219,4238	5.7434,9117	7.6122,5504	10.0626,5689	13.2676,7847	17.4494,0227	30
30½	4.4286,7307	5.9132,8609	7.8741,7784	10.4574,1981	13.8518,6300	18.3010,8749	30½
31	4.5380,3949	6.0881,0064	8.1451,1290	10.8676,6944	14.4617,6953	19.1943,4250	31
31½	4.6501,0673	6.2680,8325	8.4253,7029	11.2940,1339	15.0985,3067	20.1311,9624	31½
32	4.7649,4147	6.4533,8668	8.7152,7080	11.7370,8300	15.7633,2879	21.1137,7675	32
32½	4.8826,1206	6.6441,6825	9.0151,4621	12.1975,3447	16.4573,9843	22.1443,1587	32½
33	5.0031,8854	6.8405,8988	9.3253,3975	12.6760,4963	17.1820,2838	23.2251,5442	33
33½	5.1267,4267	7.0428,1834	9.6462,0645	13.1733,3722	17.9385,6429	24.3587,4745	33½
34	5.2533,4797	7.2510,2528	9.9781,1354	13.6901,3361	18.7284,1093	25.5476,6986	34
34½	5.3830,7980	7.4653,8744	10.3214,4090	14.2272,0420	19.5530,3508	26.7946,2220	34½
35	5.5160,1537	7.6860,8679	10.6765,8148	14.7853,4429	20.4139,6792	28.1024,3685	35
35½	5.6522,3379	7.9133,1069	11.0439,4176	15.3653,8054	21.3128,0824	29.4740,8442	35½
36	5.7918,1614	8.1472,5200	11.4239,4219	15.9681,7184	22.2512,2503	30.9126,8053	36
36½	5.9348,4548	8.3881,0933	11.8170,1768	16.5946,1098	23.2309,6098	32.4214,9286	36½
37	6.0814,0694	8.6360,8712	12.2236,1814	17.2456,2558	24.2538,3528	34.0039,4859	37
37½	6.2315,8776	8.8913,9589	12.6442,0892	17.9221,7986	25.3217,4747	35.6636,4214	37½
38	6.3854,7729	9.1542,5235	13.0792,7141	18.6252,7563	26.4366,8046	37.4043,4344	38
38½	6.5431,6714	9.4248,7965	13.5293,0354	19.3559,5425	27.6007,0474	39.2300,0636	38½
39	6.7047,5115	9.7035,0749	13.9948,2041	20.1152,9768	28.8159,8170	41.1447,7779	39
39½	6.8703,2550	9.9903,7243	14.4763,5479	20.9044,3059	30.0847,6816	43.1530,0699	39½
40	7.0399,8871	10.2857,1794	14.9744,5784	21.7245,2150	31.4094,2005	45.2592,5557	40
40½	7.2138,4178	10.5897,9477	15.4896,9963	22.5767,8503	32.7923,9730	47.4683,0769	40½
41	7.3919,8815	10.9028,6101	16.0226,6989	23.4624,8322	34.2362,6786	49.7851,8112	41
41½	7.5745,3386	11.2251,8246	16.5739,7860	24.3829,2784	35.7437,1305	52.2151,3846	41½
42	7.7615,8755	11.5570,3267	17.1442,5678	25.3394,8187	37.3175,3197	54.7636,9924	42
42½	7.9532,6056	11.8986,9340	17.7341,5710	26.3335,6206	38.9606,4723	57.4366,5231	42½
43	8.1496,6693	12.2504,5463	18.3443,5475	27.3666,4042	40.6761,0984	60.2400,6916	43
43½	8.3509,2359	12.6126,1501	18.9755,4810	28.4402,4703	42.4671,0548	63.1803,1754	43½
44	8.5571,5028	12.9854,8191	19.6284,5959	29.5559,7166	44.3369,5973	66.2640,7608	44
44½	8.7684,6976	13.3693,7191	20.3038,3647	30.7154,6679	46.2891,4497	69.4983,4929	44½
45	8.9850,0779	13.7646,1083	21.0024,5176	31.9204,4939	48.3272,8610	72.8904,8369	45
45½	9.2068,9325	14.1715,3422	21.7251,0592	33.1727,0413	50.4551,6802	76.4481,8422	45½
46	9.4342,5818	14.5904,8748	22.4726,2338	34.4740,8534	52.6767,4185	80.1795,3205	46
46½	9.6672,3792	15.0218,2628	23.2458,6237	35.8265,2046	54.9961,3314	84.0930,0265	46½
47	9.9059,7109	15.4659,1673	24.0457,0702	37.2320,1217	57.4176,4862	88.1974,8526	47
47½	10.1505,9981	15.9231,3585	24.8730,7274	38.6926,4210	59.9457,8513	92.5023,0291	47½
48	10.4012,6965	16.3938,7173	25.7289,0651	40.2105,7314	62.5852,3700	97.0172,3878	48
48½	10.6581,2980	16.8785,2400	26.6141,8783	41.7880,5347	65.3409,0579	101.7525,3320	48½
49	10.9213,3313	17.3775,0403	27.5299,2997	43.4274,1899	68.2179,0838	106.7189,5716	49
49½	11.1910,3629	17.8912,3545	28.4771,8098	45.1310,9775	71.2215,8731	111.9277,8652	49½
50	11.4673,9978	18.4201,5427	29.4570,2506	46.9016,1251	74.3575,2008	117.3908,5288	50

FIG 8 Pages from John Smart's *Tables of Interest*, 1726

Courtesy of Benjamin Wardhaugh

North, (Baron North by courtesy—his father was both Baron North and Earl of Guildford). The book's travels continued, and it ended up in the possession of Edwin Bayford's Family Mourning Warehouse, Barnsley, in the later nineteenth century.

It's a large slim volume, and one of the early owners had it bound with attractive marbled endpapers. It was evidently of interest to a wide range of different individuals over a long period of time: Henry Metcalfe corrected some errors in the text, and the Barnsley firm valued it enough to stamp it with their name and address. For a firm of drapers and mourning outfitters its value must surely have been practical rather than antiquarian. William Inwood, architect and surveyor, based his mathematical tables of 1811 in part on Smart's, it seems, thus ensuring their indirect survival during the nineteenth century through the nearly two dozen editions of *Inwood's Tables of Interest and Mortality*.

Smart himself seems to have been more of a gentleman than a tradesman. Although he was responsible for four previous, simpler volumes of monetary tables, he didn't publish anything else except a minor topographical work on the wards and parishes of London. He certainly seems to have been outside the class that would have sold books or instruments or run a mathematical school. He is therefore slightly unusual among authors of practical mathematical works. His tables of interest were dedicated to the governor and directors of the Bank of England—this during the period when that institution was still suffering psychologically if not practically from the events of 1720—and this personal connection may explain his interest in the subject.

One might expect Smart's tables, then, to be a gentlemanly production of little practical value. Yet they were in fact resolutely functional in character, and this particular copy shows every sign of having passed among a series of individuals to whom it was of use rather than ornament. The contents ranged across several different, fairly basic, monetary situations: the reduction of shillings, pence, and farthings to decimal parts of a pound; the calculation of simple and compound interest; the value of annuities and life insurance; and reverse calculations aimed at finding a fair purchase price for a given annuity or insurance.

It was not very distant in character from the kind of thing to be found in many almanacs of the period, but it was set out at much greater length, with finer gradations and a wider range of prices represented, so as to enable more accurate and rapid handling of specific situations. For example, Moore's almanac in this period gave a table of interest at 5 per cent and contained just under 200 entries. Smart's table of interest had over 4,000 entries, included eleven different interest rates, and covered periods of time ranging from a single day to twenty-five years. Although the pages were large, the print was small, and a lot of information was packed into this slender book.

The book was prefaced by a brief explanation of the use of decimals. As '100 is One hundred', so '.01 is One hundredth Part', and so on, together with instructions and examples for performing arithmetical operations on decimal fractions. Evidently Smart envisaged his book being used by people for whom these elements would be new. (Decimal arithmetic still tended to be introduced separately from 'vulgar' arithmetic, either in a separate section of an arithmetic primer or even in a different volume altogether, making it possible for a student to have worked through a fairly comprehensive course on arithmetic without seeing any decimal fractions.) For the further assistance of beginners, problems were set out at various points in the book, and space was given on the final page where their solutions could be noted down (no early owner of this copy did so).

At the end of the book Smart gave a short history of English money and of interest rates, concluding with some remarks on the phenomena of inflation and monetary fluctuation:

> there is likewise *a natural Interest of Money*, which may be very well compared to the Market Price of *other* Commodities; . . . I look upon Money it self to be a Commodity, which like others rises and falls as there is a Demand for it.

So the book offered instruction and even speculation as well as practical usefulness: a very brief course in what money was as well as a rather longer one in how to handle it. It was aimed, perhaps, at the junior bank clerk keen to learn more of the profession. The untypical story of this particular copy illustrates the interest that such matters aroused in eighteenth-century Britain, where mathematical tables were an increasingly pervasive tool of practical life.

Imagine a barrel whose ends are formed by two *different* ellipses. So suggested John Dougharty, writing in *The General Gauger* in 1707. 'Gauging' was the art of measuring, calculating, or estimating the volumes of vessels of various kinds, usually for the purpose of taxing their contents. Monsters like Dougharty's misshapen barrels might not have had much to do with the real life of a tradesman or a customs officer but his manual, dedicated to the seven 'Chief Commissioners and Governors' of the English excise and intended for the 'Edification of young Officers' in that profession, remained in print for four decades. Gauging books were another part of the growing genre of writing on practical mathematics in the eighteenth century. They—and books on many other forms of weighing, measuring, and calculating—taught those who made a living from such practical uses of mathematics, whether at mathematical schools or through private study.

Gauging was an old profession. In principle, the problem of assessing quantities of taxable goods was virtually as old as taxation itself, but the growth both of international trade and of the complexity of tax regimes gave it a particular importance in seventeenth- and eighteenth-century Britain.

Gaugers weren't popular. Back in 1676, *Poor Robin's Intelligence* had made a gauger's 'measuring a Vessel' a metaphor for sexual misconduct ('the goods were Prohibited'), in a passage which reinforced that familiar sense of numerate or geometrical activities as inherently dishonest. There, he also hinted at the tendency of gaugers to aspire to a larger smattering of Euclid than their 'Mechanick

Education' gave them any right to. Gauging was, indeed, a profession with both a reputation for unsavoury dishonesty and some aspiration to be a demonstrative science rather than just a manual practice. Although there was no 'worshipful company' of gaugers (or indeed of any kind of mathematical practitioner), the profession employed a significant number of people, and was sometimes presented in print as a way for a lad with a facility for numbers to 'get on'—the 1753 title page of *Problems in Practical Geometry* aimed that book particularly at 'those who aim at getting into the Excise', for example.

A result was that there were any number of different teach-yourself manuals for gauging on the market at any one time. At least fifty were published between 1700 and 1750, including *The Royal Gauger, The Complete Gauger, The Complete Gauger Improved, Every Man his own Gauger*, as well as several official manuals issued by the customs and excise. There were even popular songs about the subject:

> There is a thing they Name Excise,
> 'Tis known by all that Swallow,
> And that, and drink will oft Disguise,
> The Man they call good Fellow,
> The Excise then is got,
> By drinking [of] your Pot,
> And there are those I lay Wagers,
> Are more Knaves than Fools,
> For they walk by your Rules,
> And some People call 'em the Gaugers.

Thus the anonymous author of 'The Seizure, or the Sack and Gauger', printed in 1719 (it went 'To the Tune of The Dragon of Wantly', apparently), a ballad in which three customs officers received the nasty surprise of finding a stolen corpse in a sack they searched over-zealously.

In essence, a gauger's manual consisted of a list of rules for estimating the volumes of liquids contained in differently shaped vessels in different situations. Only rarely could a real cask or tun be usefully approximated by the obvious, dull cylinder. If the sides bulged outwards a more plausible, if not always a more accurate guess, made it a portion of an ellipsoid (the shape produced by rotating an ellipse about its long axis). If they bulged a lot, the cask might be considered a portion of a sphere. Much of the art of gauging was in choosing which geometrical shape to use as an approximation to the barrel at hand: the popular options were the shapes formed by truncating and joining cones, ellipsoids, and paraboloids. More geometrically impressive, but surely of less use in practice, were situations where the ends of the barrel were not circles but ellipses, squares or even parallelograms. A final class of cases, of real practical concern, addressed the situation where a partly filled barrel—even one of quite a simple shape—was tilted so that its ends were no longer parallel to the surface of the liquid it contained.

Here is Dougharty's rule for this last situation, when a tun in the shape of a truncated cone is tilted and then filled until the contents just touch the circular top. Take the square of the diameter of the top (in inches); add half the product of the diameter of the top with the diameter of the bottom; multiply by the vertical distance between the highest point of the tun and the level of its contents; and then divide by a conversion factor which, to give a result in gallons, is 1077.

Dougharty, like many mathematical writers, also taught. In fact, he ran a mathematical school in Worcestershire as well as working as a gauger and surveyor. His book was fairly successful, and part of its appeal was surely its reliance on formulaic procedures like this one, together with lots of worked examples. Such a procedure could be memorized just like the rule of three, and the underlying geometry need not be understood by the student.

Here is another example, taken from the instructions on gauging appended to the best-selling general textbook *The Young*

Mathematician's Guide, whose author, John Ward, was himself a highly successful professional gauger: 'chief surveyor and gauger general in the excise', in his own words. His account of the subject reached as far as a cask in the form of a 'truncated spheroid', considered in the case when it was laid on its side and filled to a depth of a given number of inches. A full geometrical solution to such a problem would have been appallingly difficult. Ward's rule was this:

> To Twice the Area of the Bung Circle [the base], add the Area of the Head Circle [the top]; multiply their Sum into one Third of the Length, and the Product will be the Content.

Rules of this kind, simple and usable though they were, were all too easy for one author to copy from another. Title pages would claim that the material was set out according to a new plan, better organized or more clearly presented, but these were all too often empty boasts. By the eighteenth century the market in gauging books was sufficiently full that authors who aimed at more than modest sales needed to offer something more.

That 'more' typically took the form of more mathematics. Understandably, Dougharty assumed that his readers' ability to follow, and their patience with, complex geometrical proofs was rather limited, and avoided giving anything like a complete geometrical justification for rules like the ones quoted above. But the trend was away from such moderation. Dougharty's book came with a mathematical introduction setting out, from the beginning, the basics of decimal arithmetic and the small amount of geometry needed to understand and use the rules in the main part of the book. But others, including Ward's, came with much more elaborate mathematical preambles justifying their supposedly uniquely accurate rules, with many dozens of pages of geometry set out in a Euclidean style with theorems, proofs, lemmas, and corollaries.

This tendency was evident by the end of the seventeenth century when William Hunt, another expert gauger, set out the following theorem concerning the 'cylindroid', a more general case of the misshapen cask mentioned above: 'When the Bases are both Elliptical, but unequal, and Disproportional, or Inverted; Or if one Base be a Circle, and the other an Ellipsis, the tun is called a *Cylindroeid*.' This beast was to be gauged using Hunt's Theorem 130, which set out a rule to use in such cases. Take the longest diameter of the bottom; add half the longest diameter of the top, then multiply by the shortest diameter of the bottom. Next, take the longest diameter of the top, add half the longest diameter of the bottom, and multiply by the shortest diameter of the top. Multiply these two results together, and divide by a given factor depending on what units are being used; then, finally, multiply by the height of the barrel to find its volume.

Hunt was no eccentric, but a man at the centre of his profession. One of his books carried a commendatory poem by the astrologer, mathematician, and almanac-maker Henry Coley, asserting that:

> Both in the Theory and the Practick part
> Of Gauging, which is now become so pure
> A *ne plus ultra* may be put I'm sure.

Yet Hunt did not in fact represent the culmination either for gauging or for this particular tendency towards intellectualizing and incomprehensibility. By the middle of the eighteenth century his successors would be discussing almost every variety of cask shapes produced by rotating parabolas, hyperbolas, and circles about an axis, even giving instructions for finding the volume of distillation vessels formed by piecing together sections of several such shapes. The use of calculus—William Hunt was a pioneer in this particular respect—further widened the range of possibilities, but at an awful cost in terms of mathematical complexity.

We may well ask what the function of this kind of thing was, particularly when it was aimed supposedly at 'the meanest capacity', or at least at those who had no direct interest in the geometric proofs behind the simple rules they used, even if they could have understood them. Few gauging manuals were models of lucidity in their more mathematically dense parts, and after perusing them one starts to suspect that the incomprehensibility is part of the point, the elaborate diagrams and remorseless geometrizing being a sort of 'proof by intimidation', before which the reader was expected to assent and be convinced, but not really to understand. This use of mathematical exposition was intended not so much to instruct as to impress—a kind of mathematical self-advertisement meant to distinguish a gauging book by its sheer complexity and sophistication.

All things have their limit, and the readers and users of these books were far from naïve. Up to a point, a more complex and fuller geometrical justification might have given one more confidence in the rules in which it culminated, and might have led one to purchase *The General Gauger* rather than *Gauging Perfected*, therefore rewarding the writer (and the publisher) who adopted such a strategy. But taken to extremes, mere mathematical verbiage surely did not carry much conviction, and the bulk of some of the more elaborate gauging books must have reduced their value as pocket-friendly practical manuals to consult on the job. Reaction, perhaps inevitably, set in.

That reaction could take the form of books which emphasized the rough-and-ready rules which were 'presently used' by real customs officers, or of manuals which gave as much emphasis to the processes of professional qualification involved in becoming an excise officer as to mathematics: one even offered '*authentic* Forms of such *Certificates, Petitions, Oaths,* &c as are requisite'. Other books took the term 'gauging' to include all the practices of measuring, weighing, and even surveying, so as to present a general handbook on the mathematics and practice of measurement, rather than a specialized and overcomplex account of liquid measures alone.

Reaction could also take the form of a minimal approach to the presentation of the rules of gauging. From the early eighteenth century or before, some almanacs included short instructions for gauging casks. William Turner, for example, compiling an almanac 'for the centre or middle of England' in 1701, included a one-page table 'whereby to gauge or find the quantity that any Vessel or Cask contains'. The table had just two columns, and came with a single paragraph explaining its use. It was, in effect, a ready reckoner for squaring a number and multiplying it by a conversion factor. All you had to do was look up the results of that calculation for the top and bottom diameters of your barrel, add them together (taking the second one twice), and multiply by the height. A very similar procedure to some set out by Dougharty and his colleagues, but presented in the simplest possible form.

Naturally, such a very simple presentation came at some cost in accuracy. In particular it discarded the multitude of different cases, relating to slightly different shapes of barrel, discussed by the larger manuals. But for many practical situations it would give results that were accurate enough for use, since a few fluid ounces here or there might be accounted for by wastage in any event.

Like the interest table as a strategy for avoiding calculation, such a minimal presentation of gauging amounted to a strategy for avoiding both geometry and more complex arithmetic. Despite the attractions to authors of geometrical complexity as a form of self-display, in order to make a real connection with the lives of working people it was sometimes far better to take a simpler approach and condense mathematical operations into the form of ready reckoners or of instruments.

Another very widespread mathematical practice was surveying. Indeed—like many practical mathematicians—several of the gaugers we met in the previous section also worked as surveyors. Figure 9 shows the 'plane table' made in the 1750s for John Thompson, a surveyor, and owned and used by him until the end of his life twenty years later.

FIG 9 John Thompson's plane table, by George Adams, 1756–60
© Museum of the History of Science, Oxford

Like many other instruments from the period it was signed by its maker: 'Made by G* ADAMS in Fleet Street London, Inst: Makr: to his Royal Highness the Prince of Wales'. It was therefore made after 1756, when George Adams gained that appointment, and before 1760, when the Prince became King George III. A museum curator once called it 'the finest plane-table we have ever seen'. The wood of the table is cracking a little due to its age, but this instrument retains the beauty of old wood- and brass-work, and the sense that it tells a story of hard practical use in its day.

George Adams was one of the premier makers of scientific instruments in Georgian England. After an apprenticeship beginning in 1724, Adams made and sold instruments from his shop in London from the 1730s until his death in 1772. The family business continued another four decades and more, until 1817. His clients included the Mathematical School at Christ's Hospital, the Prince of Wales, George II, and George III.

Part of Adams' output aimed to fulfil gentlemanly interest in mathematical and scientific devices as recreations or amusements. He was responsible, for example, for making natural philosophy instruments for lectures and demonstrations at the king's private observatory at Kew. His *Micrographia Illustrata* was in some respects a tribute to Robert Hooke's great *Micrographia* of the previous century, lavishly illustrated and setting out for the curious what could be seen with (Adams') microscopes. It went to four editions during Adams' lifetime, 'for the Assistance of those, who are desirous of surveying the extensive Beauties of the minute Creation'. Later in his career he sold similar toys: apparatus for performing demonstrations with static electricity, air pumps, and such ready made experimental demonstrations as 'A gun-lock for striking a flint in vacuo' or 'A water pump, to prove there is no such thing as suction'.

But these more spectacular instruments were by no means the whole of his work. His own catalogue of the instruments he made and sold around the 1760s shows a broad trade in instruments for practical use. For example:

Plain tables, with an index and sights
Ditto larger, better made, and stronger
Ditto, with telescopic sights

Or, again, 'Plain gunner's quadrants', or 'Gen. Williamson's instrument for howitzers, mortars, &c'. Plainness and strength are the characteristics of instruments destined for demanding practical use. Adams made demonstration instruments for the Royal Military Academy at Woolwich, and supplied instruments to the East India Company and for James Cook's second voyage in 1771. The records of the Office of Ordnance show that he supplied it with more than 1,500 instruments during his career, with a total value of nearly £2,500. His instruments travelled widely as a result, particularly to the Mediterranean and to North America. John Millburn, the historian of the Adams firm, quotes a typical bill for his instruments:

> To George Adams the Sum of Eighteen pounds six shillings & six pense for the Plain Table undermentioned &c. by him delivered at the Tower for use of Major Green at Gibraltar per Warrant dated 27 April 1761 and certificate.

The specific items supplied were the plane table itself, 'with a Box', a 'Theodolite Second Best' and surveying chains of 100 feet and 50 feet. The plane table would have been similar to the one pictured in Figure 9.

Adams was both a maker of traditional types of instrument and an innovator, an inventor of new designs. His novelties included a new kind of quadrant for finding the altitude of the sun at sea, a new design of refracting telescope, a 'trigonometrical octant' to supplement the work of quadrants, and 'Adams's protracting parallel rule, for drawing parallel lines at any given distance, or given inclination, to another line'.

What about John Thompson, the surveyor for whom this instrument was made? Born in the 1720s, he hailed from Leicestershire, and was involved as a surveyor in fenland drainage projects around the 1770s, when he may have used the instrument depicted in Figure 9. He wasn't a published writer, but we know something about him because a collection of material relating to his and his son's work has found its way into the Museum of the History of Science in Oxford, including instruments and survey books.

As well as the plane table and tripod made for him by George Adams—said to be to Thompson's own design—Thompson also owned a telescopic level made by another great instrument maker, Thomas Wright. Evidently the Thompsons were sufficiently successful to afford instruments made, even tailor-made, by the very best.

What exactly was a plane table? It was one of the basic instruments for surveying land, or at least for surveying relatively small areas (techniques for surveying on the scale of a whole country remained diverse, doubtful, and inventive—a title page of 1716 referred intriguingly to 'the Learned Mr *Whiston's* and Mr *Ditton's* New Method of Surveying *England* by *Explosions*', a method employing the speed of sound to estimate distances). A plane table would be used for surveys of perhaps part of a large estate: a few fields or so.

In essence, a plane table was simply a flat board mounted on a stand, together with a device for sighting: either a rule, or in more advanced instruments a telescopic sight. In its later forms certain refinements could be added, such as scales on the sighting rule or on the brass square surrounding the wooden board itself. However, these tended to detract from the simplicity and therefore the advantages of the board in its plainer form.

The board had to be oriented using a compass and levelled, usually with a spirit level. Once it had been set up, to use it the surveyor would attach a sheet of paper to the board, using pins or a clamp, and mark a suitable starting point on the paper. Then the sighting rule would be used to draw lines of sight on the paper from that starting point to the various features of the land which were to appear in the survey.

Once that was done, plane table and surveyor would move off along one of the lines of sight, measuring the distance to the feature at the end of it using a chain or another measuring device. The distance would be transferred, at an appropriate scale, to the paper, thus giving the position on the survey of the surveyor's new location. There, the plane table would be set up and used again to find a new set of sight lines to all the features of the landscape that were of interest.

In principle, the intersection of the two sets of sight lines on the paper would give accurate positions for all of those features.

In practice, greater accuracy might be attained by moving plane table and surveyor to a third location, or by checking some or all of the distances in the survey by direct measurement.

One of the important features of the plane table was that it allowed a survey to be constructed directly by sighting, without the need to measure any angles or, necessarily, to measure more than one length. It thus minimized some of the possibilities for errors to enter a survey, and minimized the degree of mathematical knowledge needed by its operator, although in practice many surveyors were skilled mathematical practitioners in their own right. Once again, then, this was a strategy for avoiding the need to do more mathematics than was necessary: this time one which obviated the need for geometry or trigonometry.

> If any Gentleman or other, especially Ladies, that desire to look into their disbursements, or layings out, and yet have not time to practise in Numbers, they may from Mr. *Humfrey Adamson* dwelling near *Turn-stile* in *Holburn* have those incomparable Instruments, that will shew them to ply Addition and Substraction in l. s. d. and whole numbers, without pen, ink, or help of memory.

I couldn't leave the subject of useful mathematics and its instruments without mentioning this wonderful advertisement from 1674 by Nicholas Stephenson, a gunner at the Tower of London. The instrument in question was an early example of a mechanical adding machine, this one devised by Sir Samuel Morland, inventor and natural philosopher.

Such machines had their origin with the French mathematician and philosopher Blaise Pascal—he had presented one to Queen Christina of Sweden, which Morland almost certainly saw during his travels as a diplomat. Morland built his own machines during the 1660s and presented them to Charles II. One was simply an adding device: it had three dials for use in currency calculations—divided into, respectively, four, twelve, and twenty parts for pennies, shillings, and guineas—and five more for decimal calculations. A system of gearings and stop-pins enabled the dials to be used to add or subtract numbers with, as Morland put it in his own publications, no need to rely on pen, paper or memory, and—in theory—no risk of making mistakes. It measured only about four inches by three, and was just a quarter of an inch thick: roughly the size of a pocket almanac. Unfortunately its usefulness was sharply limited by the absence of a carrying mechanism.

A later, rather larger machine by Morland could also perform multiplication, and a final device employed broadly similar means to assist the performance of trigonometric calculations. These provided only a rather limited degree of simplification compared with other ways of doing the calculations: Pepys found Morland's machines 'very pretty, but not very useful'. One was seen, and disparaged, by Robert Hooke in January 1672: 'very silly', he called it, and immediately set about to design one of his own.

The late seventeenth and early eighteenth centuries saw a burgeoning of the uses of mathematics and of the profession of the mathematical practitioner: an expert in one or more of those uses. Practices like gauging and surveying were professions, and individuals expert in those branches of practical mathematics would very often possess other related skills too. Some ran mathematical schools, and others wrote about the useful mathematics they knew and used.

Also on the increase was the frequency and sophistication with which mathematics was used—or was expected to be used—by those who would not have called themselves mathematical practitioners. In any number of fields a new optimism about what mathematics could do for you resulted in the production of books, techniques, tables, and instruments that were supposed to be 'absolutely necessary' for one kind of workman or another. In 1738 Batty Langley, for example, published

> The Builders Compleat Assistant, or a Library of Arts and Sciences, Absolutely Necessary to be understood by Builders and Workmen in general.

The arts and sciences in question were arithmetic, geometry, architecture, mensuration, plane trigonometry, surveying, 'mechanick powers', and hydrostatics. The compilers of almanacs responded to the same trend, inserting into their publications mathematical information intended to be of practical use. Tables of weights and measures, tables for working out interest or wages, or brief rules for weighing and measuring: we can imagine a reader, perhaps of Hatch's generation, struggling through such practical calculations based on only the rules and tables given in the almanac, and in the process slowly becoming more comfortable with, and more comprehending of, basic arithmetical operations and their practical usefulness.

Occasionally almanacs showed a more explicit interest in mathematics and its usefulness. John Wing's almanac *Olympia domata* contained a discussion of 'The Use and Excellency of Mathematical Learning', which continued over several issues in the late 1710s. The writer found uses for mathematics across the whole of human activity, arguing that it made one more useful to king, country, and indeed oneself, and lamented the scarcity of those who truly understood it. English mathematical education was compared unfavourably with that of the French, and the writer remarked of the mathematical school at Christ's hospital that ''tis to be wished, there were many more such'. For this writer, as for many at this time, 'Mathematical Learning' was a single thing, not a set of distinct practices to be learned separately as needed, but an 'immediately useful' general toolkit which deserved to be much more widely available.

Poor Robin himself adapted to changing times in similar ways. His almanac never declared its usefulness on its title page, but during the second decade of the eighteenth century it began to include new material on useful mathematics in the form of tables of weights and measures: how many ounces in a pound, and so forth. His almanacs also began to include tables of the tides, with explanations of their use. The author seems to have felt some embarrassment, at least initially, at this change from buffoonery to seriousness. Into the list of weights and measures he inserted a laboured pun about four legs and fore-legs, while the unsmiling explanation of the use of the tide table was followed without interruption by the whimsical remark that 'The Sea is the stable of Horse Fishes, the Stall of Kine-fishes, the Sty of Hog-fishes, and the Kennel of Dog-fishes'.

However, over time the tables of interest and of the value of investments became an established part of Poor Robin's almanac, and they expanded to include more sophisticated information, such as this:

> A Table shewing how many Years Purchase a Lease or Annuity . . . for any Number of Years under 30, is worth presently, at Interest upon Interest, at Six in the Hundred, and shewing also how to discount any Lease in being, and the true Value of the Reversion after any Number of Years.

The point of this particular table was to show the potential buyer or seller of a given annuity how much it was worth at present. Suppose, for instance, I own an annuity that will pay £10 a year for the next eighteen years, and I wish to sell it to raise ready cash. How much is it worth? Assuming the rate of inflation used in Robin's table—6 per cent—it will be worth '10 Years, 9 Months, and 9 ten parts of a Month's Purchase': that is, a little more than £107. Strikingly sober by Robin's earlier standards, this could almost be an extract from a manual of accounting (and it very probably was—there is no reason to suppose that Poor Robin's author calculated this table personally).

In later years there were tables of bread weight and tables of wages, and in the 1730s Poor Robin even succumbed to including topographical information: a table of the sizes, populations, and positions of the English counties. Such dry stuff was common in other almanacs, but he had previously avoided it. The tide table expanded over the years, from its original half dozen towns to nearly fifty.

A later addition was a ready reckoner to find the price of 112 pounds (a 'great hundred') of a commodity given the price of one pound:—one year Robin called it a 'Ready Rhino'. Others were tables for the conversion of money from one currency to another and, 'for Noveltys Sake', a table showing how to tell the time at night from the rising and setting of the Pleiades. In keeping with the new emphasis on use and practicality, Robin's stance towards astrology was at the same time becoming less jolly and more stern. Astrologers were now routinely 'ass-trologers', their practice a cheat rather than an amusement.

But the early eighteenth century was not the best time for Poor Robin. Early in the century his author felt confident and secure enough to claim in the yearly preface that he troubled to continue writing solely in order to reward his loyal fans. But by 1740 matters had deteriorated to the point that the (presumably different) author could sign himself 'Poor, Poor, Poor, Poor Robin' and remark that

> His *Ægyptian* Task-masters they so keep him under,
> That if he should crack a Joke 'tis a Wonder.

By now, he claimed, he was struggling to make ends meet, to pay the rent, to afford even beer. The claim to be writing only for the public good now had a pathetic, not a pompous ring:

> We barter our Pleasure for Posthumous Breath,
> And lose many Meals to live after Death.

It's not clear just how much recognition from posterity could really have been expected for Poor Robin's almanac, although it continued to sell upwards of 10,000 copies a year, and to make a profit.

By the middle of the century Robin was both aware of himself as the bearer of quite a long heritage (title pages noted that such and such an issue was the eighty-eighth, and so on), but also attempting to reposition himself so as to retain a share in the changing almanac market. Almanacs were by no means the only cheap way of obtaining practical mathematical information, and Poor Robin's jokes were by now very old ones. The introduction to the 1750 issue claimed that 'we hope to serve you, not helter-skelter, slap-dash, confusedly topsyturvy, as in the Days of Yore' but 'diligently enough, carefully and manifestly'. Poor Robin was becoming middle aged, perhaps.

Over time the proportion of new material appearing in his almanac each year fell—more and more sections were reused from year to year. Poor Robin retained his sharply royalist line, pointedly omitting any reference to the Commonwealth in his potted history of Great Britain and retaining the twofold calendar that had been a trademark since the beginning. Loyal Britain never lacked opponents for Robin to mock in the comical, 'fanatical' calendar for the year—the Stuart Pretenders were obvious targets for much of the eighteenth century—but the mockery eventually became somewhat more measured. The specific title-page reference to papists was dropped, and criticism of Protestant dissent, too, vanished almost entirely. The calendar of mock saints was reduced, in the end, almost entirely to figures from fiction and folklore—'sad Sinners, and sorry Saints', as the title page put it—together with the odd bit of whimsy from Robin's own brain (the nineteenth of February, 1765: Saint Pancake).

Despite all this, Poor Robin was a popular enough figure to receive the sincere flattery of at least two imitators during the first half of the eighteenth century. The *Rhode-Island Almanack*, calculated for Newport but intended for use across the whole of New England, purported to be written by him. According to its title page, the 1728 issue was 'the first ever published for that Meridian'. It also seems to have been Poor Robin's last excursion to Rhode Island, although the same almanac continued in print, attributed to other authors, for some years. The printer was named in the book but not the real author:

> Tho' I have not given you my *proper Name*, yet I assure you I have had one the greatest part of half an hundred Years; and I know of no Necessity for parting with it at this Time.

It was in fact James Franklin, Benjamin Franklin's brother. He made no attempt to imitate the particular style of the original *Poor Robin*. His astrological predictions were mostly concerned with the weather (the sixteenth of March: 'Rain, with wind, this Week'), and he supplemented these with simple rules in the tradition of natural astrology for foretelling the weather from the sky and the behaviour of livestock ('If Sheep do bleat, play or skip wantonly, it is a Sign of wet Weather'). The twelve-month calendar was supplemented by three brief lists: of eclipses, of Quakers' meetings, and Baptists' meetings. What fun the old Poor Robin would have made of that.

A decade or so later another American styled himself 'Poor Robin', this time in Philadelphia. In its later issues this almanac was entitled *Poor Robin's Spare Hours* and specialized in medical information and moralizing tales. The issue for 1755 contained a short essay on the value of alms-giving and the frame of mind and practical precautions which should attend it, a brief fable in verse concerning the evils of vanity and envy, a poem on the subject of true happiness and how to find it, and recipes for medicines to cure both cholera and cancer. This remarkable information was interspersed with the astronomical calendar, following it there were further sections about the movements of the heavens and the tides, and there was a fascinating collection of local data including the times and places of court sessions, religious meetings and fairs, and descriptions of the highways and roads between major towns. On the last page a Philadelphia brushmaker advertised his willingness to buy hogs' bristles at good prices. Almanacs had lost nothing of their heady complexity in the New World, although once again the particular flavour of Poor Robin the wise fool was absent.

Also in Philadelphia, and a still clearer descendant of Poor Robin, there appeared from the 1730s *Poor Richard's Almanack*. It was alleged to be written by one 'Richard Saunders', but the author was in fact none other than Benjamin Franklin himself. The 1758 preface gathered 'Poor Richard's' advice and aphorisms as the wildly successful (and still well-known) *Way to Wealth*. It would ultimately be reprinted in well over 1,000 editions and become a key piece of American writing from its period. As late as 1832 a further imitator, *Poor Richard's Journal*, would start up in London, containing legal and medical advice for its readers.

Franklin stated in his autobiography that he wanted his almanac to be 'a proper Vehicle for conveying Instruction among the common People, who bought scarce any other Books'. Poor Richard was unmistakably a new draught upon the Poor Robin tradition ('creditors are a superstitious sect, great observers of set days and times'), and he showed what riches that tradition still held.

Chapter 5

'Close and demonstrative reasoning'

BEAUTIFYING THE MIND

In 1770, or thereabouts, Master Thomas Porcher took up his pen in Walworth, near London, and began to write in a fat, hardbacked volume:

> Geometry (a Word formed from the Greek... to *Measure*) is the Science of Extension & considers Lines Surfaces and Solids; for all Extension is distinguished into Length Breadth and Thickness.

Rather more than 200 pages of close, neat handwriting later, he concluded on the final page of the book with the fantastically careful drawing show in Figure 10, illustrating 'The fundamental Projection of the Series of Sines Tangents and Secants, on the Plane Scale'.

Thomas was born in 1756. We know very little about his family situation—his geometry book speaks of moderate affluence, because it suggests private tuition on a fairly substantial scale, but we do not know what career or profession Thomas was being prepared for. He married a Kate Allen, and the couple apparently remained in London until the end of their lives. Their descendants live on to this day. (One of their sons seems to have made his fortune in the New World. By 1811 a Samuel Porcher was the owner of a plantation in South Carolina. He named his son, Thomas, after his

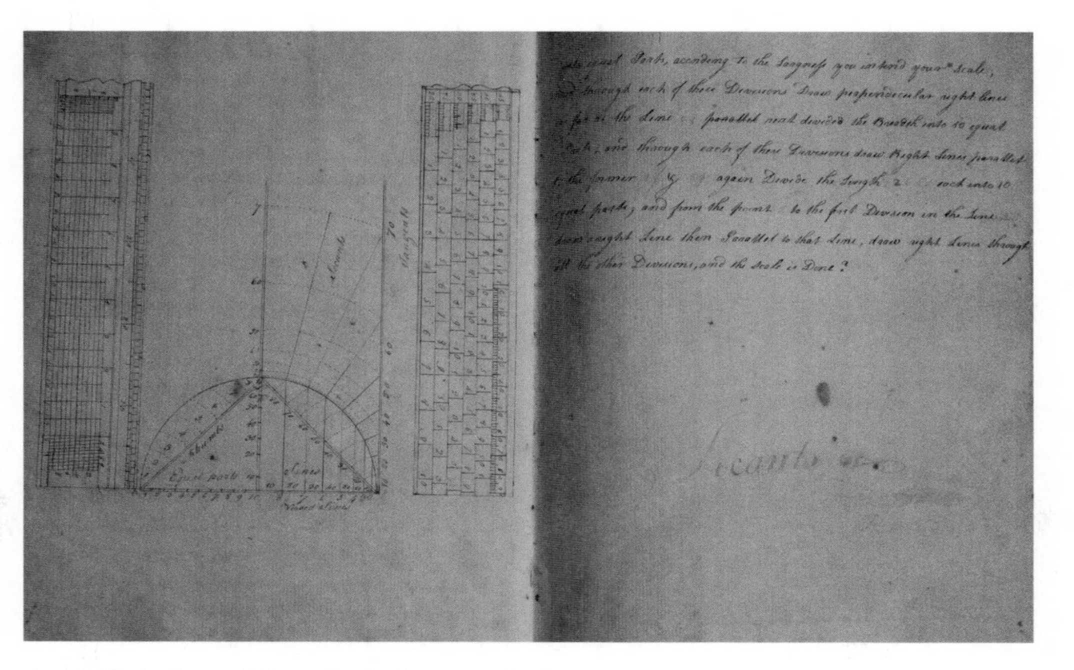
[illegible] equal Parts, according to the Longness you intend your Scale,
[illegible] through each of these Divisions Draw perpendicular right lines
[illegible] the Line [illegible] parallel next divide the Breadth into 10 equal
Parts, and through each of these Divisions draw Right Lines parallel
to the former [illegible] again Divide the Length [illegible] each into 10
equal parts, and from the point to the first Division in the Line
draw aright Line then Parallel to that Line, draw right Lines through
all the other Divisions, and the Scale is Done?

FIG 10 The final page of Thomas Porcher's geometry book, *c.* 1770
Courtesy of Benjamin Wardhaugh

father, and the plantation, Walworth, after his family's home in England. The family remained in the area throughout the nineteenth century, and a Porcher Avenue in nearby Eutawville was named after them.)

Thomas followed a course of geometry that contained a lot from Euclid's *Elements*. It began, after a very brief preface about the nature of geometry, with a series of definitions and axioms, some of which were paraphrases of the ancient Greek master. After these came an ordered sequence of propositions and proofs: many were Euclidean, but many were not, or were very loose paraphrases. All this material was organized into five 'parts', whose contents sometimes mirrored the organization of the *Elements*: part five, for example, covered proportional lines, as does Euclid's fifth book. The course also covered the drawing of lines, the construction of plane figures, and the 'inscription' and 'circumscription' of figures. A sixth section contained 'geometrical theorems', and ended with the proof of Pythagoras' theorem, which famously concludes book one of the *Elements*.

That took Thomas about half-way through his book. In the second half he learned to apply the geometrical results obtained so far to practical ends: the measuring of surfaces and of solids, and plane trigonometry. The latter was presented with its own set of axioms and examples, elaborately divided up on occasion into different cases, each with its 'construction' and its 'calculation'. At the very end of the book a few of the examples made reference to real-world problems, with lovely pictures like that shown in Figure 11: the distance of the castle from the tree was to be determined using the height of the castle and the angle it made when seen from the foot of the tree.

Thomas Porcher was plainly not making all of this up, nor does his geometry book look at all like notes he could have taken in a classroom. It's just possible that they were dictated off the cuff by a very able teacher, but the high level of organization and the

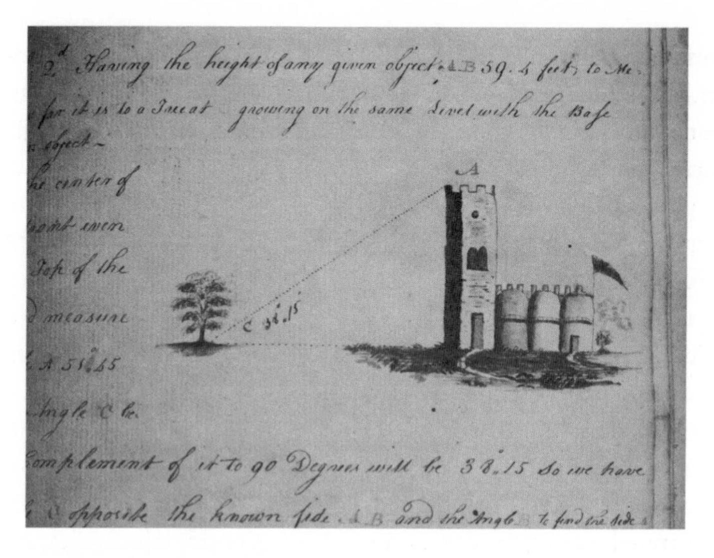

FIG 11 Geometry meets reality, in Thomas Porcher's geometry book
Courtesy of Benjamin Wardhaugh

unfaltering clarity of expression suggest rather strongly that Porcher was in fact copying all of this out from a written source or sources. There were a lot of geometry textbooks on the market in the second half of the eighteenth century, and rather than copying out one of them verbatim—for what would have been the point of that, except to wear out pens and fingers?—he was perhaps piecing together different sources or at least supplementing a printed book. In places there are miniscule annotations which may have been intended to give the names of the authorities used for particular sections of the text, although they are scarcely legible now. Or—and perhaps this is more likely—he could have been copying a book that only existed in the form of the teacher and author's own manuscript.

Whatever the details of its sources, Thomas Porcher's book is distinguished by the extraordinary beauty of its handwritten pages. By complete contrast with the unruly scrawl of an arithmetic exercise book like that of Isaac Hatch—and, it must be said, with some geometry exercise books from the same period—it was made with a quite remarkable care. Corrections are rare, ugly crossings-out almost unknown; the handwriting is small and even throughout a text that must have taken dozens of hours to write out. We might almost suspect that Thomas Porcher was the teacher, not the pupil, but for the fact that he calls himself 'Master' Thomas on the decorated title page.

In addition to the text itself, which occupies the right-hand side of every two-page spread, there are diagrams—dozens and dozens of them—drawn with a fastidious neatness which echoes that of the handwritten text. Red ink, black ink, pencil, and watercolours (red, yellow, and grey) were all employed in order to make it quite clear just what was going on when each definition or proof was illustrated. Facing the definition of 'Curves or Curvilinear figures', for instance, are examples of a circle, an ellipse, ovals, a spiral and 'An irregular curvilinear figure' which looks like a child's drawing of a cloud, the more so as it was coloured in using grey watercolour. As well as the colouring, each was outlined in black ink and labelled in red.

All of this is quite overwhelming when it is repeated over so many pages, with never a wobble in either the geometrical lines themselves or the consistency of their presentation. In the final few pages of the book, with their illustrations of trees and castles, there is the real sense of a young boy at last letting his imagination have some play. That imagination had first been disciplined through a very long apprenticeship of lines, angles, and planes.

And surely that was much of the point of this book: not just to beautify the page but to beautify the mind. It was virtually a commonplace in the eighteenth century that the study of geometry disciplined not just the hand but also the mind, making it fit for higher studies by training it in the habits of clarity and concentration. The tight, fastidious structure of the material Porcher was asked to copy down, and its emphasis on precise verbal and diagrammatic expression, on thematic and of logical structure, speak, surely, of a concern to instil virtuous and ennobling qualities of thought, as well as virtuous habits in the handling of pen, brush, and pencil.

A less laborious way to produce a beautiful volume somewhat like Porcher's—and thus to beautify the mind of a pupil—is illustrated by *Problems in Practical Geometry...for the Use of Schools, And the Instruction of those who aim at getting into the Excise,* by W. Brown, teacher of mathematics. One copy never entered a school and was used to teach someone who probably had no intention at all of getting into the excise, or into any trade at all. This was Cosmas Nevill of Holt, a minor member of one of the noble families of England (the Nevills held the title of Baron Bergavenny and, later in the century, Earls of Abergavenny).

As a child in Leicestershire his mathematics tutor was one Samuel Davis (he shares his name with a mathematical fellow of the Royal Society and servant of the East India Company, a somewhat younger man and a specialist in Indian astronomy, but we do not know for certain of a connection between the two). Davis bought an unbound copy of *Problems in Practical Geometry* for a shilling shortly after its publication in 1753. Instead of having it bound in the normal way, he proceeded to insert about twenty additional sheets, on which were neatly handwritten a series of carefully planned supplements to the book. Only then was the whole bound, in such a way that this additional material would appear here and there throughout the book.

Davis' additions were written in two colours of ink with very great care. They provided extra worked examples, extra geometrical theorems, additional explanations, and better proofs. Davis also wrote some short explanations and elucidations on the printed pages themselves. The result was a unique volume, a composite of print and manuscript which was almost as much his work as that of

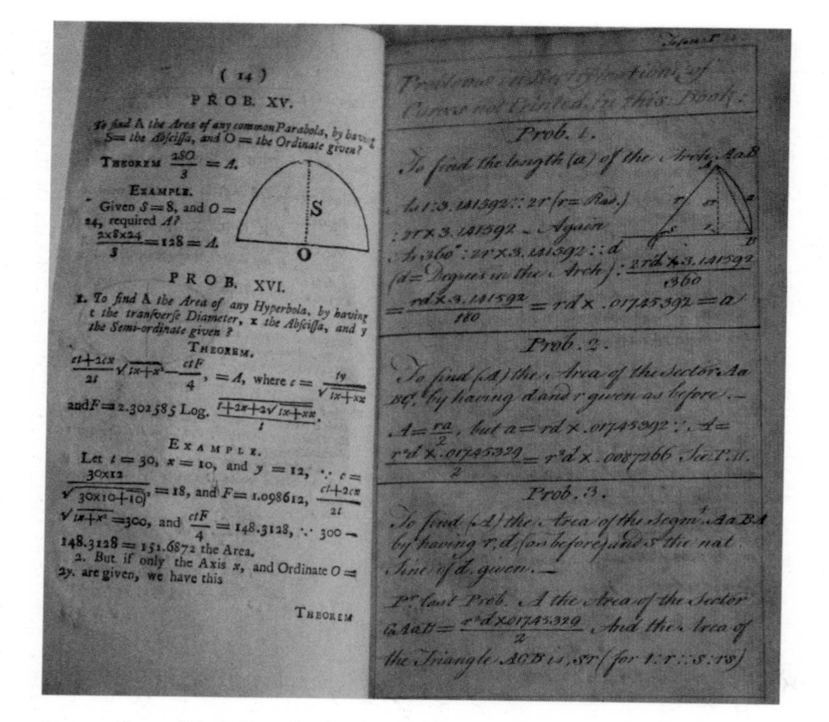

(14)

PROB. XV.

To find A *the Area of any common Parabola, by having* S = *the Abscissa, and* O = *the Ordinate given?*

THEOREM $\frac{2SO}{3} = A$.

EXAMPLE.

Given $S = 8$, and $O = 24$, required A?

$\frac{2 \times 8 \times 24}{3} = 128 = A$.

PROB. XVI.

1. *To find* A *the Area of any Hyperbola, by having* t *the transverse Diameter,* x *the Abscissa, and* y *the Semi-ordinate given?*

THEOREM.

$\frac{ct + 2cx}{2t}\sqrt{tx + x^2} - \frac{ctF}{4} = A$, where $c = \frac{ty}{\sqrt{tx + xx}}$

and $F = 2.302585$ Log. $\frac{t + 2x + 2\sqrt{tx + xx}}{t}$.

EXAMPLE.

Let $t = 30$, $x = 10$, and $y = 12$, $\therefore c = \frac{30 \times 12}{\sqrt{30 \times 10 + 10^2}} = 18$, and $F = 1.098612$, $\frac{ct + 2cx}{2t}\sqrt{tx + x^2} = 300$, and $\frac{ctF}{4} = 148.3128$, $\therefore 300 - 148.3128 = 151.6872$ the Area.

2. But if only the Axis x, and Ordinate $O = 2y$, are given, we have this

THEOREM

Prob. 1.

To find the length (a) of the Arch AaB

As 1 : 3.141592 :: 2r (r = Rad.) : 2r × 3.141592 . Again

As 360° : 2r × 3.141592 :: d (d = Degrees in the Arch) : $\frac{2rd \times 3.141592}{360}$

$= \frac{rd \times 3.141592}{180} = rd \times .01745392 = a$

Prob. 2.

To find (A) the Area of the Sector AaBC, by having d and r given as before.

$A = \frac{ra}{2}$, but $a = rd \times .01745392$ ∴ $A = \frac{r^2d \times .01745329}{2} = r^2d \times .0087266$

Prob. 3.

To find (A) the Area of the Segm.t AaBA by having r, d (as before) and s the nat. Sine of d. given.

Pr. last Prob. A the Area of the Sector CAaB $= \frac{r^2d \times .01745329}{2}$. And the Area of the Triangle ACB is, sr (for 1 : r :: s : rs)

FIG 12 Samuel Davis' supplements to *Problems in Practical Geometry*, c. 1753

Courtesy of Benjamin Wardhaugh

the author named on the title page, and indeed he wrote in his own name on the title page of this copy, as though to assert his rights over what it contained.

The contents of this book began roughly where Thomas Porcher's left off, with fairly advanced problems and theorems relating to plane measurement: it would have come not at the beginning but perhaps towards the end of a school-level course on mathematics. As with Porcher's book, the lucid organization of the page seems to have been important, although here the material to be studied was mostly of a sturdily practical kind. By the end of the book the discussion had reached such matters as 'cask-gauging' and other parts of the excise officer's art. But Davis, by his additions, made it a geometry textbook of a slightly more elevated kind, where the elegance of the proofs and the completeness of the logical organization were just as important as the utility of the results.

We can imagine Davis' dilemma, and how he struggled to reconcile his own ideas about how the geometry of measurement should be taught with the practical angle he found in the available printed accounts of that subject. He wished, perhaps, to extend an able pupil with some more advanced material about solid geometry, but the only textbooks he could find that were less difficult than the Greeks themselves were stolid practical manuals for aspiring tradesmen. This, too, in the context of teaching a pupil whose use of the material probably had much more to do with beautifying his mind than with getting him into the excise business. His solution was ingenious, and he succeeded in creating a volume which turned W. Brown's *Problems* into something rather different.

During the early eighteenth century 'Automathes', a young boy, was stranded due to a series of misadventures on a desert island with little more than a few books and a selection of mathematical instruments. During several years of solitude he nevertheless achieved remarkable feats of self-education. Since he had not yet learned to read when he was cast away, the geometry book and the instruments were his first choice when the heroic effort began:

> had it not been for the Cuts interspersed here-and-there, especially in the History, and the Geometrical Schemes, which abounded in the Mathematical Book, I believe I should have for ever laid them aside, without farther Examination. But these, being more intelligible than the Letters, convinced me of their Design, and were afterwards a Direction for me to hold the Book. And the Mathematical Volume became of great Use to instruct me in the Principles of that Science, though without the least Knowledge of a Letter contained in it.

He learned the use of ruler, compasses, and 'Brass Semicircle', and thus various geometrical operations: 'to measure a right Line by equal Parts, to describe a Circle, to bisect a Right Line'. By similar means, arithmetic unfolded itself to him:

> I had also a frequent Sight of the nine Digits or Figures, both in the Book and upon the Instruments; and, observing the same Figures always to denote the same Numbers, I became at length perfect[ly] acquainted with their Use; and likewise discovered how these alone, with the Character called a Cypher, were adapted to express all Numbers imaginable, by the Variation of their Places.

So, although letters continued to elude him, being 'put to signify Things I have not the least Notion of', he was able to acquire a wide range of mathematical knowledge. Having reached Pythagoras' theorem, he extended his studies to solid bodies and conic sections, aided both by 'the actual Formation of several regular Bodies out of Clay' and by the pictures in his book. He studied basic number theory, too. And his mathematics did not remain merely speculative: he constructed a large sundial, a meridian line, and a quadrant, and achieved some knowledge of the motions of the heavenly bodies, all of which helped him to keep track of time on the island and to deduce something of its location.

Automathes was eventually rescued, allowing him not just to tell his remarkable story, but to compare the mathematics he had learned with that of the outside world. Though he found some discrepancies in, tantalizingly, 'a Sort of Arithmetic peculiar to myself', for the most part everything he knew turned out to be part of ordinary mathematics, capable of being located in—usually—Euclid's *Elements*.

This intriguing tale (fictional, I'm sad to say) demonstrates clearly how mathematics could become a metaphor for universal certainty and mental improvement. For Automathes it remarkably extended his mind's reach beyond the narrow circle of his island, so as to make contact—of a kind—with the rest of the world. Like a mathematical Robinson Crusoe (and Crusoe himself had salvaged some mathematical instruments from his own shipwreck, remarking on the possibility of teaching oneself 'every mechanick Art' by beginning with mathematical reason), his journey in physical and intellectual solitude was in the end an edifying one.

'Automathes' was alone, but he wasn't unique. An ancient story told of Aristippus, a Socratic philosopher who, thrown on the shore of Rhodes by shipwreck, came across elaborate geometrical diagrams drawn in the sand of the beach. On seeing this he called to his companions to take heart, because he had seen the traces of mankind: humanity would certainly not be wanting in people who cultivated these arts. In some versions of the story he went further and inferred the possibility of lucrative employment for a philosopher from the apparent cultivation of geometry.

The eighteenth century produced a wealth of stories about imaginary travels and imaginary places: hapless travellers cast up on sun-kissed beaches meeting ideal communities of nudists, republicans, or talking chickens, according to preference. The idea of mathematics as a line connecting the traveller back to reality was used by several of their authors, although as we saw in Chapter 2, the fictional consequences were not always desirable ones.

So, as the examples of Thomas Porcher and of Automathes show, learning geometry was a special experience. Geometry was eternal and universal—it was uniquely accessible to the human intellect and it bore a unique relationship with truth, certainty, and humanity. The universality, certainty, and reliability of geometry had growing practical importance in a world where the uses of mathematics were constantly expanding. But perhaps its most important effect was not on the objective world, but on the human mind.

Geometry as an improver of the mind thus found a place—as well as in fiction—in the literature on the 'usefulness of mathematics' which stretched from the inaugural addresses of university professors to serialized discussions in popular almanacs. We met one example, from an almanac, in the last chapter. Isaac Barrow, the first Lucasian Professor at Cambridge, lectured on the subject, Christopher Wren ended his inaugural lecture as Gresham Professor of Astronomy with a discussion of mathematics as the paradigm of certainty, and Nicholas Saunderson discussed the usefulness of mathematics in his inaugural Lucasian lecture.

At a somewhat more popular level, the Scottish physician and satirist John Arbuthnot wrote on the usefulness of mathematics in an essay which was reprinted throughout the first half of the eighteenth century. 'In the search of truth, an imitation of the method of the *Geometers* will carry a Man further than all the *Dialectical* rules'. (A distant reflection, possibly, of Descartes' belief that geometrical truths were uniquely 'clear and distinct' to the mind.) For those who wished to discipline the mind and fit it for higher studies, there was no better tool than geometry: it could 'charm the passions, restrain the impetuosity of imagination, and purge the Mind from error and prejudice', 'accustoming it to *attention*', and 'giving it a habit of *close* and *demonstrative reasoning*'. A century later Byron would allude to the same qualities when he said of Madame de Staël that 'The want of a mathematical education, which might have served as a ballast to steady and help her into the port of reason, was always visible'.

William Leybourne's popular collection of mathematical recreations *Pleasure with Profit* of 1694 was introduced with similar claims, in this case for mathematics as a virtuous and improving leisure activity, superior to almost all other kinds of mental or physical exercise. Mathematics, for Leybourne, had as its object 'no less than the *whole World*'. Its demonstrations were 'as infallible as *Truth* it self... it is one of the most *Excellent Sciences* in *Nature*', and therefore 'it may best become the Industry of *Man*, who is one of the best *Works* of *Nature*'. Indeed, he claimed, mathematics actually underlay the most valuable physical exercises:

> Play at the *Billiard Table* hath not its Peer: It exercises the whole *Body moderately*; the *strength* of the *Arm judiciously*: It *directs* the *Hand Geometrically*, and the *Eye Optically*: For the attaining to be an Exquisite Proficient in *playing* at it, depends wholly upon putting in *practice* that *Axiome* of *Euclid* in his *Catoptiques*; which *demonstrates*, that, *The Angles of Incidence and Reflection are ever more equal.*

Such views of mathematics and geometry tended, like Leybourne, to give place of honour to Euclid as exemplifier of these characteristics of logical method and sturdy deduction. The *Elements* had first been translated into English in 1570, in a version which remained the only English one until 1651, but between then and the end of the seventeenth century there were five more English Euclids, and by the end of the eighteenth century there were a dozen. Euclid's visibility was rapidly increasing, and the Euclidean text itself, sometimes attractively simplified, amplified, or rearranged, was accessible to many more people than previously. The *Elements* did, of course, employ logical deduction to a high degree, and many of its propositions consisted of requests to prove particular assertions, as in the famous story about Thomas Hobbes' introduction to geometry:

> He was forty years old before he looked on geometry; which happened accidentally. Being in a gentleman's library Euclid's Elements lay open, and 'twas the forty-seventh proposition in the first book. He read the proposition. 'By G—,' said he, 'this is impossible!' So he reads the

> demonstration of it, which referred him back to such a proof; which referred him back to another, which he also read. *Et sic deinceps*, that at last he was demonstratively convinced of that truth. This made him in love with geometry.

But much of the contents of the *Elements* is expressed in the language of mechanical operations with compass and rule, and many of its propositions request not proofs but 'constructions'. For example, the first proposition of Book 1 asks the student 'To construct an equilateral triangle on a given finite straight line'. Geometry was thus a discipline not just of the mind but of the hand as well.

Another oft-repeated story illustrates the concentration that this combination of mind and hand might ideally produce. Archimedes was,

> as fate would have it, intent upon working out some problem by a diagram, and having fixed his mind alike and his eyes upon the subject of his speculation, he never noticed the incursion of the Romans, nor that the city was taken. In this transport of study and contemplation, a soldier, unexpectedly coming up to him, commanded him to follow to Marcellus; which he declining to do before he had worked out his problem to a demonstration, the soldier, enraged, drew his sword and ran him through.

Here, in Plutarch's account of the death of Archimedes, geometry was an almost meditative, absorbing practice making the geometer indifferent even to the most extreme vicissitudes. The pursuit of such indifference continued—albeit in a smaller way—in the early modern period. Samuel Johnson was helped to cope with being stranded on the island of Col by reading part of James Gregory's *Geometry*, on which he 'made some geometrical notes in the end of his pocket-book'.

A result was that geometry was frequently called a more gentlemanly study than arithmetic. Patterns of education could reflect that, with geometry taught to the sons of gentlemen and arithmetic and the rule of three to those of lower status. But exceptions were common, too. In Chapter 4 we saw how some authors of gauging books used a quasi-Euclidean presentation in an attempt to increase the status and the saleability of what were essentially practical manuals. The trade of surveying likewise—in some presentations—employed Euclidean geometry. On the other hand, the eighteenth century saw various attempts to bring arithmetic and algebra into the sphere of self-improving gentility as polite accomplishments.

In Friendship two Sisters together reside,
With Virtue replete; each a Stranger to Pride:
Maria for Beauty with Venus may vie,
And Cloe for Wisdom Minerva defy;
Maria is prudent in ev'ry Degree,
Whilst Cloe is court'ous, good-natur'd and free.
From what's underwritten their Ages I ask:
Resolve it, dear Ladies; nor think't a hard Task?

Given

$$x^2 + xy + y^2 = 1087,$$
$$x^4 + x^3y^3 + y^4 = 45777295;$$

To find the Value of x the Age of *Maria*, and *that* of y the Age of *Cloe*.

The *Ladies' Diary* (or 'Woman's Almanack'), begun by a Coventry schoolmaster, John Tipper, in 1704, appeared annually until 1841. It thus had a similar lifespan to *Poor Robin*, and it took the same pocket-sized format as an almanac. It had some of the same content too: a yearly calendar with anniversaries and holy days, and a list of eclipses (it gave eclipse computations in detail). But it always had its intended audience of intellectually curious women in mind, and endeavoured to provide material which would be of interest: recipes, sketches of the lives of notable women, and so on. It had no prognostications, nor even weather predictions.

During the *Diary*'s early years the focus shifted decisively. A contribution of arithmetical problems by a (male) reader proved so popular, and similar material in subsequent years generated such a response, that mathematical problems and puzzles soon pushed aside much of the other content. Within a few years the *Diary* had come to serve a highly distinctive purpose, facilitating the discussion of mathematics by readers—both male and female—keen to improve their minds. The content of the *Diary*—apart from the calendar—came to consist quite substantially of mathematical problems, together with verbal 'enigmas' and other puzzles, and of course the solutions to last year's problems sent in by readers (there were prizes, taking the form of free copies of the *Diary* for a certain number of years).

This was certainly congenial to Tipper, who taught at his school:

> *Writing, Arithmetick, Geometry, Trigonometry*; the *Doctrine* of the *Sphere, Astronomy, Algebra*, with their Dependents, *viz. Surveying, Gauging, Dialling, Navigation*, and all other *Mathematical Sciences*; Also the True Grounds and Reasons of *Musick*.

Later editors of the *Diary* included well-known practical mathematicians and mathematical writers. In the later eighteenth century it was for a long period edited by Charles Hutton, whom we will meet in Chapter 7: one of the most significant individuals for popular and practical British mathematics in the period.

The *Diary* found imitators, including the inevitable *Gentleman's Diary*, and there were several compendia, indexes, and supplements to its contents, including the *Palladium; or Appendix to the Ladies' Diary* which was published annually during the third quarter of the century. It was the pattern for, in all, more than a dozen mathematical periodicals intended for a broad audience which appeared during the eighteenth century, though most were very short-lived.

The number of women contributors actually dropped over the 140 years of the *Diary*'s existence, and on the whole many more men than women seem to have contributed problems and solutions. Yet the *Diary* remained committed to attracting female readers and as far as possible women contributors, and its combination of this emphasis on a female readership with a good deal of mathematical content proved very successful. Sales peaked at around 30,000 a year in the middle of the century, and only a handful of the many other almanacs of the eighteenth century did nearly so well. Its problems found their way into other almanacs, textbooks, and teaching materials (Robert Gardner wrote out what looks very much like a *Diary* problem in his 1775 'Book of Acompts').

The chance to see one's mathematical problems or solutions in print clearly appealed to both female and male readers of the *Diary*. An early editor wrote of receiving four or five hundred letters from women alone in response to the *Diary*'s puzzles and challenges (it is not quite clear over what period), at a time when six or seven thousand copies of the *Diary* were selling each year. One admirer wrote in to praise what it was doing for mathematics: 'the diary', he wrote, 'has incited and led many persons to the study of mathematics, who otherwise perhaps would not have turned their thoughts that way'.

More than that, it made possible a distinctive community, even a kind of mathematical club. By writing in to the *Diary*, readers became participants in a small but long-lived world of intellectual exchange. The problems and solutions had something of the tone of those in *The Mathematician*, which we met in Chapter 3.

But while the first editor intended in principle to provide 'something to suit all conditions, qualities and humours', it was in fact the leisured rather than the working classes who came to make up this mathematical community. It was therefore a rather different institution from a club like the Spitalfields Mathematical Society, with a rather different function: the improvement of the mind not for practical use but for the mind's own sake.

The *Diary*, indeed, contained a characteristically polite mode of mathematical discussion—one in which modesty and self-deprecation were rather important. A high proportion of contributors adopted pseudonyms (sometimes quite fantastic ones: Blowzabella, Chrononmononpublicus, Birchoverensis). Thus, although the *Diary* embodies the public self-presentation of a certain group of eighteenth-century women—and its attention to the intellectual life of women is rare for its period—it often shows those women adopting conventional and even self-effacing roles. As a matter of policy, names were printed only if the correspondent specifically requested it, and as many as 90 per cent of the women who corresponded with the early editors on mathematical subjects did not as a result see their names printed in the *Diary*. The title page of the *Diary* praised the intellectual accomplishments of the (British) women it represented:

> Hail! happy Ladies of the *British* Isle,
> On whom the Graces and the Muses smile...
> Nature to make your *Triumph* more complete
> To peerless Charm has added piercing Wit.

But editors had the peculiar task of praising women's mathematical abilities while directing a publication in which those abilities were rendered all but invisible.

A fictional dialogue of 1755, whose aim was to teach natural philosophy to the reader by means of a brother-and-sister dialogue, indicated the same problem in its introductory pages:

> I often wish it did not look quite so masculine for a Woman to talk of Philosophy in Company... how happy will be the Age when Ladies may modestly pretend to Knowledge, and appear learned without Singularity and Affectation!

One manifestation of decorousness and restraint was that for the first few decades problems and even solutions for the *Diary* were required to be given in verse, 'which', according to Tipper, 'will be still the more taking among the ladies'. The requirement distinctly cramped the expression of mathematical ideas as well as the self-presentation of the contributors. A later example illustrates the problem, and one way around it:

> Dear ladies fair, I pray declare,
> In Dia's page next year,
> When first it was I 'gan to pass
> My time upon this sphere.
> My age so clear; the first o'th'year,
> In years, in months, and days,
> With ease you'll find, by what's subjoyn'd,
> Exact the same displays.

A footnote gave the actual mathematical content of the problem:

> $xy + z = 238$
> $xz + y = 158$
> $x + y + z = 39$
> Where x = the years, y = the months, and z = the days of my age, the first of January, 1795.

In the same year another piece of doggerel verse gave a far-from-trivial question in three-dimensional geometry in terms of the form of a petticoat. The problem amounted to an advanced piece of gauging, in fact: the volume of a section of an ellipsoid, given certain of its measurements. The *Diary* did not avoid subject matter from the practical world:

> Three ships, A, B, and C, sailed from a certain port in north latitude, until they arrived at three different ports, all lying under the equinoctial; A sailed on a direct course, between the south and the west 175.62 leagues; C sailed 133 leagues between the south and east; and B sailed a course between A and C 102 leagues, making the angle or rhumb with A, equal the angle that C made with the equinoctial. Hence it is required to find the port sailed from, each ship's course, and distance from each other, and their respective ports? And to solve it by an equation not higher than a quadratic?
> (By Mr. Robert Heath, 1737)

Such questions as these brought the mathematics of trade and craft into the *Diary*'s sphere of polite discussion. The historian of the

Diary, Shelly Costa, notes that this apparent paradox was of a piece with the emergence in eighteenth-century Britain of an openly commercial kind of civility, a politeness unashamed of its commercial nature:

> Unperturbed by the low status that still tinged such studies... the *Diary*'s correspondents embraced 'mechanical' contexts, translating them through polite discourse and formal mathematics to make them both palatable and entertaining.

Money, land, and artisanal practices such as gauging and surveying recurred as settings for the *Diary*'s mathematical questions, and the *Diary*, writes Costa, 'imbued the mathematics of commerce with the values of gentility'. In this juxtaposition, as she explains, 'the abstract mathematical techniques enjoyed by the *Ladies' Diary*'s readers were posited as superior to the trade methods of artisans'. One contributor illustrated this attitude very neatly in her solution to a problem involving a tinker:

> Well, bonny brave tinker, to save thee from ruin,
> The kind British lasses are active and doing.

At the same time another change was gradually taking place. Tipper's initial desire had been to 'please' with fairly trivial mathematical riddles, rather than to 'puzzle' ('If to my Age there added be / One Half, one Third, and three times Three; / Six score and ten the Sum you'll see, / Pray find out what my Age may be'). But readers and later editors pushed the *Diary* in a much more challenging, less playful direction—one that was intellectually exclusive and ultimately less suited to polite discussion:

> The Greatest *Spheroid*, and Parab'lic *Conoid*,
> Inscrib'd in a *Cone* are by Art,
> From whence as *below*, the Contents you're to show,
> Of each separate Solid apart.
> *Diam. of Cone's Base* = 35 *Inches, and its Altitude* = 30 *Inches.*
> (by Mr. James Terey of Portsmouth, 1752)

By the later part of the eighteenth century the problems were far from easy, and they covered the whole circle of the mathematical sciences. A compilation of its mathematical problems produced in 1817, towards the end of the *Diary*'s existence, identified twenty-five categories of 'mathematical learning' represented in its problems, from harmonics, algebra, and astronomy to hydrostatics, navigation, and pneumatics. The algebra and geometry required were often difficult, and calculus made fairly regular appearances too.

Thus a typical issue might contain as its problems two or three abstruse constructions in plane geometry, the same number of purely algebraic questions, plus a few such delights as this:

> I have seen a sheep leap from a bridge very high, into water, and swim out. Now, if a globe, whose weight is 112 pounds, and one foot in diameter, fall from an eminence ten yards high, how deep must the water be, just to destroy all the globe's velocity, supposing the density of air, water, and the globe to be as the numbers $1/5$, 1000, and 10,000 respectively.
> (By Mr Thomas Milner, Lartington Free School, 1798)

The reader who could follow all of this would have been well-educated indeed. Ultimately the balance between mathematical seriousness and accessibility to polite discussion—particularly at a time when women's education was failing to keep pace with men's—proved too difficult to sustain. The *Ladies' Diary* merged with the *Gentlemen's Diary* in 1841 and ceased publication in 1871.

But the *Diary* at its best—when numbers of women were prepared to provide solutions and to contribute their own mathematical questions—provides evidence of a body of mathematically educated females who are not prominent in other historical sources. (Although they have left traces here and there: subscription lists for mathematical works in eighteenth-century Britain include the names of over a thousand women.) It illustrates, too, the long-lasting although ultimately passing idea that mathematics—with all its benefits to the mind—might become a usual polite accomplishment.

For a period in the 1760s and 1770s, *Poor Robin* made contact with the tradition of the *Ladies' Diary* through the person of Thomas Peat. Peat was one of the founders of *The Gentleman's Diary*, an imitator of and rival to the *Ladies' Diary*, and he was its sole editor for a quarter of a century until his death in 1780. He was also, during roughly the same period, the editor of *Poor Robin*.

He was a versatile man: a mathematical practitioner whose activities ranged from lecturing on natural philosophy to valuing timber, land surveying, and drawing up house plans, according to his own advertisements. His lectures claimed to cover mechanics, hydrostatics, optics, pneumatics, astronomy, and 'the use of the globes', and to include all the experiments needed to explain both natural philosophy in general, the properties of matter, and the 'Laws by which it acts'. In addition he ran a school:

> Youth genteelly boarded, and carefully instructed in Writing, Arithmetic, Drawing, and the most *useful* Parts of the *Mathematics*, on reasonable Rates.

This remarkable man was based in the north of England—he lived for a time in Leicestershire, and his lectures were given in Nottingham.

Peat, as might have been expected, brought a new vigour to *Poor Robin*, and he took an interest in its readers' intellectual development. We don't know exactly when he took up the editorship, but the issue for 1765, for example, carried a discussion of taxation under the title 'Considerations on the present high Prices of Provisions, and the Necessaries of Life', which seems characteristic of a new direction for Robin. As always, Poor Robin was on the side of the poor: he argued here that the present high prices of such necessaries as beer could not be justified by their cost of manufacture, and provided a brief calculation to show that beer 'may be brewed by private Families for 17 Shillings per Barrel', little more than a third of the price you would pay in a pub.

The same issue advised financial prudence to those in a position to invest their money:

Covet all, loose all;
Therefore Poor Robin advises all
  To be merry and wise,
And lend their Money on *good* Securities.
Happy the Man who is content,
Tho' he can make but *Four per Cent*
His Fortune in the Funds to fix,
And trusts no private Hands for Six.

More impressive still as evidence of Robin's new-found seriousness was an exposition of the motions of the planets in the Copernican system of astronomy. It appeared in several sections across consecutive annual issues of the almanac, giving the impression that Peat liked to think *Poor Robin* was not being read and thrown away but kept from year to year, the consecutive issues perhaps even being bound together to form a volume. His hopes may have been misplaced: such handling of *Poor Robin* is not much in evidence in the copies which survive.

The discussion of Copernican astronomy occupied a column on the same page as Robin's comic calendar, where it forms a very odd contrast indeed. The calendar directed the reader to the celebration of 'Buxome Nell' and 'Saint Landlord', while the astronomical section set out in some detail (the print is quite small) the history of the Copernican hypothesis, with *précis* of Copernicus' own book and of the relevant works of Galileo. Peat set out how the Copernican system could account elegantly for the observed facts—that some of the planets sometimes seemed to move backwards through the heavens, for instance—and even gave diagrams to supplement his explanations ('The understanding of this figure, though it be geometrical, yet supposes no knowledge of geometry'). He apparently had a keen sense of the pedagogical needs of his readers, perhaps sharpened in his work as a lecturer—he wrote of the need to 'render truth palpable', without 'enigmatic demonstrations'.

There were other changes. The table of the weights of bread was gone, replaced by a table showing the length of daylight for each day of the year, and Peat added a table of longitudes of major world cities to the useful sections of the almanac.

It is remarkable that the author and publisher of *Poor Robin* felt able to reposition it to this degree. (It's perhaps also remarkable that they felt it was being bought by readers who, in the 1770s, would need to have it explained to them that the earth orbits around the sun.) But *Poor Robin* was still intended for a group of readers very different from those of the *Ladies' Diary* and the *Gentlemen's Diary*, and an excess of mathematical learning could still be ridiculous for Robin. Here's his comment on those who observed the solar eclipse of 1 April 1764.

> Let sage Philosophers compute,
> And each Cœlestial Motion suit
> Their Astronomic Rules;
> What, tho' the Sun Eclips'd may be,
> Yet there's sufficient Light to see
> A Group of April Fools!

Peat's own perspective on the almanac business was seen when he dedicated his 1774 issue to his fellow authors:

> Andrews's *A Deo pendent omnia*...gives Monthly Observations enough, and (sometimes) pretty good Weather. The Gentleman's and Ladies Diaries endeavour to improve the Minds of Youth, in *Poetry*, and Mathematical Learning. Moore's *Vox Stellarum*, good Weather, and strange Hieroglyphics; Partridge's *Merlinus Liberatus*, and Definition of Astrology; Parker's Ephemeris tells us how to travel in *London* by Land and Water; Pearse affords some sublime Poetre; Saunders gives us Lectures in Mythology; Season's elaborate *Speculum Anni* produces dark Riddles and wonderful Observations; Wind mounts up to survey the Heavens; and...White's Cœlestial Atlas...is perhaps the most useful annual Publication in all *Europe*.

For Peat, the almanacs were distinguished by the useful information they contained (which could perhaps include 'good' predictions of the weather), but the prophecies and prognostications by 'Francis Moore' in *Vox Stellarum*, now in its eighth decade, were no more than 'Hieroglyphics'. The sheer diversity of that information is an indication of how far the almanac trade was still thriving.

But for all that, times were not easy for the Stationers' Company. Two rulings on the subject of copyright went against it in the late eighteenth century: in the Lords in 1774 against the claim that copyright was a perpetual property, and in the Court of King's Bench in 1798 against the claim that copyright only existed if a work was registered at Stationers' Hall. The Company's control of copyrights was much reduced, but more crucial for almanacs was a challenge by the bookseller Thomas Carnan to the Company's monopoly on their publication. In 1775 the Court of Common Pleas ruled in his favour, and within two years he was producing no fewer than thirty-one different almanacs. Other London booksellers followed with a dozen more, creating a potentially devastating situation for the Stationers' Company, which depended on the almanac trade financially.

One consequence was a relaunch of *Poor Robin* in 1777 as *Old Poor Robin*. We don't know for certain when Thomas Peat's association with the publication *Poor Robin* ended, but the continuities across the change of title were quite substantial, with the explanation of Copernican astrology running uninterrupted across the change, and I suspect it was Peat who oversaw it. The author, whoever he was, styled *Old Poor Robin* 'a new Edition of an old Almanack', and stayed in many respects close to the format and contents of his predecessor.

The two parallel calendars, serious and comic, were retained. But the new title was the occasion for a reduction in the quantity of even mock astrology that the almanac contained—the mocking prognostications at the end of each issue were reduced to just two pages. The space gained was filled for the most part with historical and moralizing material: in one year, a lengthy retelling of the legend of St Anthony of Padua.

Such polite, self-improving material would have been unthinkable in *Poor Robin* a few years earlier, typical though it was of other publications during this period. It was presumably an attempt at renewed popularity in the face of the new and intense competition for a share in the almanac trade. But the result was a 'prognostication' which had little of prognostication about it, and sadly little of the wise mockery that had distinguished Poor Robin over his previous 115 years. It's hard to avoid feeling that those who had delighted in Poor Robin's almanac in former years had been forgotten. Yet the change apparently worked: Poor Robin in his new guise continued to make a profit for the Stationers' Company for another thirty-odd years. We will pick up his story in Chapter 8, when we come to see how he finally met his end.

The Stationers' Company, meanwhile, responded to the opening up of the almanac market by campaigning for its monopoly to be restored, and after a few years the 'stamp duty' which had to be paid on all almanacs was doubled, effectively preventing the smaller competitors from continuing. Carman remained very much in business, but the Company dealt with his almanacs by buying up the rights to produce them shortly before his death in 1788. They were mostly so-called 'country almanacs', calculated for the use of particular regions. The Company would continue to produce them, and to profit by doing so, for several more decades.

The high rate of stamp duty—and it would be doubled again in 1797—prevented any new competitors from entering the almanac market, so the Stationers' Company had regained a monopoly in practice if not in principle on the above-board publishing of almanacs. But the almanac trade had become the object of a continuing and sometimes an embarrassing scrutiny. In the course of a Lords debate concerning the proposed reintroduction of the monopoly in 1779, Lord Erskine launched an attack specifically on *Old Poor Robin*, accusing it of vulgarity and even indecency. Declining out of respect for the House to quote from it, 'I know of no house', he said, 'but a brothel, that could suffer the quotation'.

It was still possible, too, for illegal almanacs to circulate, on which stamp duty had not been paid. They did so in large numbers: almost as many again as the legal ones. Some printers of illegal or pirated almanacs operated outside London, where it was harder to detect them or to act against them, while others simply sold their almanacs clandestinely in the streets of London. Unstamped versions of *Poor Robin* could go further in their satire than the official version, and thereby bring the publication and the whole business into disrepute. It was possibly to such unofficial Poor Robins that Lord Erskine had referred, since his words would be distinctly unfair as a characterization of the authentic *Poor Robin* and *Old Poor Robin* during the 1760s and 1770s.

The Stationers' Company was naturally concerned about these deleterious imitations, and Poor Robin himself periodically complained about their activities:

> If any snarling, rank mouth's Criticks dare,
> With tripple Malice, to hiss, sting, and tear
> This Work of mine, to forge, and then declame...
> Tell them I care not.

One such forger was a 'Poor Paddy', who appeared in Dublin around 1770 to produce an almanac which ran for a handful of years and claimed to be Poor Robin's 'Genuine Successor'. It focussed squarely on the weather, for which it promised 'Infallible Predictions', but something of the old Poor Robin's priorities were visible too. The printers of the almanac were named on the title page as George 'Laboissiere' and 'Oliver Alesucker', and the prognostications contained pointed advice about how to avoid being cheated: ''tis much easier to cheat a Fool out of all his Money by gaming, or an ill schemed Lawsuit, than it is to extract an Ounce of Honesty from a Knave, or of Compassion from a mercenary Lawyer'. There was even a glance at the continuing vogue for mathematical puzzles, in a question which asked the reader 'to describe, by a Perpendicular Line, in what Part of the World NO WHERE lies'.

The desire to beautify one's mind, and to use geometry or some other part of mathematics to do it, extended broadly and deeply in eighteenth-century Britain. As we have seen, Thomas Peat brought an ethic of intellectual self-improvement even into the pages of Poor Robin's almanac, a feat at which many would have balked.

Inevitably, that desire did not lack for critics. Some suggested that, far from improving the mind, excessive attention to mathematics would destroy it. Samuel Johnson's novel *Rasselas* featured a cautionary tale concerning 'one of the most learned astronomers in the world'. At first he appeared the happiest and the wisest of men: 'His comprehension is vast, his memory capacious and retentive, his discourse is methodical, and his expression clear'. But forty years of observations and calculations had strangely affected his mind, and his studies had led him to believe he controlled the motions of the planets:

> I have possessed for five years the regulation of weather, and the distribution of the seasons: the sun has listened to my dictates, and passed from tropick to tropick by my direction; the clouds, at my call, have poured their waters, and the Nile has overflowed at my command

Although he confessed that he had 'sometimes suspected myself of madness' he was slow to perceive his error. In the end, though, he passed judgement on himself:

> I have passed my time in study without experience—in the attainment of sciences which can, for the most part, be but remotely useful to mankind. I have purchased knowledge at the expense of all the common comforts of life... When I have been for a few days lost in pleasing dissipation, I am always tempted to think that my enquiries have ended in error, and that I have suffered much, and suffered it in vain.

Such disasters were not confined to fiction. One of the real-life mathematical eccentrics of the eighteenth century, William Emerson of Darlington—a highly successful writer of textbooks—cultivated such eccentricities of dress and manner that his neighbours thought him a sorcerer, believing him to have magic powers and applying to him to foretell the future for them.

But in the longer term, optimism concerning the effects of mathematical study—and, as we will see in the next chapter, the effects of mathematical organization on the world—seems to have effaced most such fears. By the end of the century claims were circulating for the therapeutic value of mathematics, even in preventing or curing mental illness. Physicians such as John Conolly, writing in 1830, would suggest the study of mathematics as part of a treatment for insanity—just as the subject was valuable for disciplining and strengthening the healthy minds of youths, it was supposed that it would have similar effects on unhealthy minds, absorbing the attention and discouraging delusion. From the same period, Wordsworth's *Prelude* contains one of the best depictions of mathematics' power to soothe, ennoble, and beautify the troubled mind.

> 'Tis told by one whom stormy waters threw,
> With fellow-sufferers by the shipwreck spared,
> Upon a desert coast, that having brought
> To land a single volume, saved by chance,
> A treatise of Geometry, he wont,
> Although of food and clothing destitute,
> And beyond common wretchedness depressed,
> To part from company and take this book
> (Then first a self-taught pupil in its truths)
> To spots remote, and draw his diagrams
> With a long staff upon the sand, and thus
> Did oft beguile his sorrow, and almost
> Forget his feeling...
> So was it then with me, and so will be
> With Poets ever. Mighty is the charm
> Of those abstractions to a mind beset
> With images and haunted by herself.

Chapter 6

'An *universal Mathesis*'

ORDERING THE WORLD

Mr Grundy will explain any difficulties in understanding the matter—so claimed Joseph Smith, gentleman scientist and correspondent of the Gentlemen's Society of Spalding, Lincolnshire. He was writing, on this occasion, about a forthcoming eclipse of the sun, to take place on 24 July 1739, and he had sent the gentlemen a whole series of detailed calculations indicating what they should expect to see. On the day itself he allowed them the use of his telescope to make observations. The letter was read and explained at the Society's meeting a few days beforehand, and we are to presume that Mr Grundy did indeed provide the required explanations: the complexity of Smith's calculations would surely have put them beyond the comprehension of many of the gentlemen present.

Spalding's society was not a specialist scientific society, it was a typical product of its age—a group of like-minded, curious, and in some cases modestly learned gentlemen who met to discuss and observe the world around them. Sometimes there was music, sometimes there were lectures. Letters would be read out from corresponding members around Britain and beyond. The Society began in 1710, and—although it lost its way for a time in the nineteenth century—it still exists.

Unlike the one in 1699, this solar eclipse was not the subject of panic in the streets, as far as we know. For the Spalding gentlemen it was an opportunity to leave the comfort of the room where the Society met and, using Mr Smith's telescope, to observe a phenomenon of nature and confirm the accuracy of his calculations. The observers quite possibly included the Mr Grundy who had taken a knowledgeable interest in the matter at the meeting on the nineteenth. The eclipse would have been observed by projecting the image of the sun onto a flat surface, so several people would have been able to watch at the same time, and we would probably not be far wrong if we imagine curious passers-by stopping to ask what was going on, and to look at the image for themselves. Or perhaps we should imagine the gentlemen seeking a secluded spot, so there wouldn't be too many passers-by.

Joseph Smith, the astronomer whose expertise—and whose telescope—the gentlemen sought, was not a member of the Spalding Society, but he seems to have been quite happy to share his knowledge and his instrument for the occasion, even if his letter did show a certain tendency to leave the gentlemen behind him. He lived in the nearby village of Fleet, and this seems to have been the only occasion when he brought his telescope—presumably a prized possession—to Spalding. But he had previously been consulted on other astronomical matters: he had sent the gentlemen his observations and calculations concerning the moon's passage in front of Jupiter and in front of the bright star Aldebaran in 1737 and, happily, those observations 'exactly answered our observations here'.

And what of Mr Grundy, who was supposed to explain Smith's letter to the Spalding gentlemen? In his forties, and regularly referred to as 'John Grundy Mathematician' in the Society's notes, he was one of those who made a living from mathematics. He had worked as a surveyor, as so many such men did (Spalding was one of the towns he surveyed, and he had presented the Society with a plan of it 'for an Ornament to their Museum'), and he was much concerned with land drainage projects, a favourite concern of the period and a subject on which he wrote for publication. He was involved with a concern gloriously named the Deeping Fen Adventurers, which existed to drain fenland and make it suitable for agriculture. Additionally, he taught mathematics at the Grammar School in Market Bosworth and, as we have seen, he seems to have functioned as a sort of mathematical consultant to the Spalding Gentlemen. A letter he wrote to them while away from the area in late 1735 gives an idea of his activities.

'In the first place I went to Chester and took a map of the Open Bay With the works that are carrying on there.' He sent his remarks, but not his map, and suggested tentatively that the Society might be interested to hear them. Next, 'I have Invented a Frustrum Hedgehog or Porcupine that will Tear away the Sands and gain a Channel in any open Bay'. He had also invented a method of 'jettying' which he believed would be more than usually robust against wind and tide. 'I have made an Improvement in the Battering Ram for Driving down Piles', reducing the number of men needed to do the work from twelve to eight while increasing both the weight of the driver and the number of strokes it could make in a given time.

As well as these highly practical matters, he had performed experiments on the cohesion of earth and clay, hoping to 'rays such Dati [data] as shall inform me in a Universal Theorem' to assist in the building of banks of earth. And, lest pure mathematics should seem to be neglected, 'I am likewise writing a Manuscript of algebra for the Benifit of my son'.

The letter might amount to a prospectus for the activities of a diligent practical mathematician in the period, yet it was also closely fitted to both the practical needs and the intellectual interests of the Spalding Gentlemen's Society, giving it a unique character and making Grundy a remarkable specimen of mathematicians' ability to survey the worlds both of practice and of intellect at this time. Nothing was done without practical purpose, yet nothing was done without a real intellectual curiosity too, and the measure of success was on the one hand the weight of the piles driven, the time taken, and the number of men used (it feels like an exercise in the rule of three, only Grundy's innovation allowed the result one would expect to be substantially improved upon). On the other hand it was the 'applause' which maps, theorems, and treatises received from 'all Parties'.

In 1738 Grundy moved with his family to Spalding, and his son became a member of the Gentlemen's Society the following year. He, too, would work as a drainage engineer, and became a founding member of the Society of Civil Engineers in 1771 (as did John Thompson, whose plane table we met in Chapter 4, and who worked with Grundy on fenland projects). The fact that the Grundys were tradesmen, and in the distinctly unglamorous trade of drainage, doesn't seem to have caused any difficulty for their enrolment as Spalding 'Gentlemen'.

The gentlemen were, in fact, rather inclusive in their membership, both socially and geographically. A correspondent, and a member of the society a few years later, was the Reverend Andreas Bing. Bing was a clergyman and a friend of certain of the more active members. He was 'Knowing in Mathematics, & Astronomy' and sent the gentlemen various astronomical observations he had made, including the track of a comet. What was unusual about him as a member of the Spalding Gentlemen's Society was that he was Norwegian by both birth and residence: his parish was in Fredrickshald in southern Norway, and his astronomical observations were made both from there and from Greenland, where he worked for a time as a missionary to the Inuit (there's a Bing's cave in Christianshaab, Greenland).

Bing was introduced to the Spalding Society by a merchant friend who exported timber from Norway to England. The friend was interested in botany, geology, and natural history—he sent specimens for the Spalding museum—but he was also prepared to translate Bing's astronomical letters from Norwegian before they were sent. Once again, the Spalding gentlemen consulted a resident expert—this time Edmund Weaver, a surveyor and almanac writer—about them, and showed their approval by making Bing a corresponding member of their Society.

Astronomy wasn't their only mathematically related activity. When the heavens were not providing much interest the Spalding gentlemen were capable of discussing such topics as the mathematics of music, the mathematics of drainage (John Grundy again), globes, fireworks, and the construction of abacuses. One member drew up a table of logarithms for the Society's use—a member later borrowed it to help him understand a remarkable book entitled *Vegetable Staticks*, which applied Newtonian methods to the study of plant growth. And the mid-1730s saw a sustained interest on the part of a few members in 'fluxions', namely the Newtonian version of differential calculus.

This interest in more learned mathematics was supported by yet another mathematical informant, of a somewhat higher professional status than John Grundy or Edmund Weaver. John Muller, born in 1688, was of German origin and worked at the ordnance office in the Tower of London and later at the Woolwich Military Academy. At the Tower one of his colleagues, William Bogdani, was a member of the Spalding Society. Muller became a member himself in 1735 and almost at once began to write to the Society on, and encourage it to discuss, mathematical subjects.

His mathematical interests ranged widely and included the mathematics of artillery as well as such purely mathematical subjects as the theory of probability, the geometry of conic sections, and 'fluxions' and 'fluents': Newtonian calculus. His book on this last subject was in preparation during his association with Bogdani: subscriptions were solicited among the Spalding gentlemen and their friends and families, and a copy was sent to their library.

Muller was also interested in finding the longitude, like so many mathematicians of the eighteenth century. But his method for doing so seems to have left him bogged down in algebra, pages of which passed from hand to hand and, again, eventually reached certain members of the Spalding Gentlemen's Society. Another problem he discussed in such writings was that of 'finding 2 Conjugat Diameters of an Ellipsis or Hyperbola containing a given Angle having any 2 conjugate Diameters given.' It may sound like a prize question in pure mathematics, but it was in fact linked with the design of burning glasses and—for one correspondent—gave rise to a discussion of Archimedes' legendary mirrors, used (supposedly) to set fire to Roman ships at Syracuse in 212 BC. Members exchanged discussions of how Newtonian fluxions worked, and on occasion took the opportunity to declare their loyalty—despite their difficulties with his methods—to the great Sir Isaac and his methods, disclaiming any intent to criticize him.

Other, rather simpler, problems and solutions by Muller were apparently sent, intended to be read out to the Spalding Gentlemen's Society at a meeting. Most members' mathematical abilities may have been modest, but they were prepared, it seems, to listen to the detailed solution of specific problems, provided they were kept short and reasonably comprehensible.

The interest provoked by Muller was in the end quite brief, despite the evident enthusiasm of some members of the Society. The Society's main interests remained in less technical scientific matters, in belles-lettres, genealogy, and antiquarianism. But through their mathematical discussions and their contact with various different mathematical informants, they witnessed and, in a small way, took part in the eighteenth century's ongoing discovery of mathematics' power to organize and control the physical world. It did so in ways that sometimes went far beyond the workaday practices of gauging, surveying, and the like, to see mathematics as a universal tool of explanation, prediction, and practical utility.

Wrote Edmund Scarburgh in his 1705 edition of Euclid, introducing the ratio and proportion material of Book 5:

> This element... stands alone as an *universal Mathesis*... equally applicable to all the Species of Quantity, to the Sciences, Geometry, and Arithmetic, and besides, universally to all other things which are

> capable of comparison, such as Force and Power in Agents, Intension and Remission in Qualities, Velocity and tardity in Motions, gravity and Levity in Ponderations, Modulation in sounds, Value and Estimation in Things, and whatsoever else may admit of any Gradation.

On such a view, the student who would master the world must master (at least) this part of Euclid, and must take at least some parts of mathematical learning very seriously.

What mathematics could do was constantly increasing during the eighteenth century. John Arbuthnot, whose essay on the *Usefulness of Mathematical Learning* we met in Chapter 5, had remarked on uses ranging from the study of animal physiology to that of the ancient philosophers, from history ('Mr. Hally has determin'd the day and hour of Julius Cæsar's Landing in Britain, from the circumstances of his relation') to 'Painting, Musick, and Architecture, which are all founded on numbers'. Allegedly, military officers were 'advanced in proportion to their skill in Mathematical Learning' while ship's carpenters were accustomed to 'implore the Geometers help'. And of course, 'it would go near to ruine the Trade of the Nation, were the easy practice of Arithmetick abolished'. For Arbuthnot, mathematics was of benefit wherever it was found, and 'what-ever some people may think of an Almanack... it is oftentimes the most useful paper that is published the same year with it'. If some of these uses of mathematics did not last, the worldview from which they arose certainly did, and it would gradually be established almost beyond question that mathematics had enormous importance for understanding the world.

For British natural science a mathematical worldview was a Newtonian one. If popular 'Newtonianism' was not always—or even often—of a very mathematical character, Newton's name did nonetheless stand as much as anything for mathematical principles, and for the idea that such principles might usefully be sought almost anywhere.

The Spalding Gentlemen's Society enjoyed an unusually wide reputation and was able to attract as members or as correspondents a range of different individuals from the world of mathematical practice and Newtonian science, ranging from Joseph Smith the local astronomer and John Grundy the surveyor, to foreign correspondents like Andreas Bing and more prominent practitioners like John Muller. It also had among its corresponding members London-based professional scientists and lecturers, including one of the most prominent popular Newtonians of the day, John Theophilus Desaguliers:

> Courses of Experimental Philosophy, and Courses of Experimental Astronomy, *publick* or *private*, in *Latin*, *French*, or *English*, are perform'd at any Time of the Year, and likewise all Parts of *pure* or *mix'd Mathematicks* taught, by the *Author*, at his House in *Channel-Row, Westminster*; where Gentlemen, who have a Mind to apply close to these Studies, may be boarded.

Thus Desaguliers, in 1728, was evidently running a boarding school in mathematics and experimental philosophy at his home in Westminster, among various other activities. He was by this time a Fellow of the Royal Society and a doctor of civil law, and he had also played an important part in the founding of the London Grand Lodge in 1717. According to his biographer Audrey Carpenter he was 'a convivial man with a gift for oratory combined with technical ability', who fitted well into an environment marked by what has been called 'the rise of public science'.

Born in La Rochelle in France, Desaguliers was the child of a French Protestant clergyman who became a refugee to England following the revocation of the Edict of Nantes. He was raised in Guernsey and London, and educated in Warwickshire and at Oxford. At Oxford he attended Newtonian lectures by John Keill, and in 1709 took over his classes after the older man's departure. Just a few years later he began to develop his reputation in London as a public lecturer and demonstrator. He rose fast, and by 1717 he was asked to lecture at Hampton Court for the King and the Prince and Princess of Wales.

In London, Desaguliers was part of a group of Huguenot refugees known as the Rainbow Coffee House group after their meeting place in Lancaster Court, off St Martin's Lane. The Huguenot community was valued in London—one of George I's earliest acts as king was to grant £15,000 annually out of the civil list for its relief, matching the sum provided by his predecessors. The Rainbow group also included the mathematician Abraham de Moivre, who apparently used to earn money at a nearby coffee house—Slaughter's in St Martin's Lane—by solving chess problems. The world of the coffee houses was famous then as now for the discussion of political, intellectual, and scientific subjects. It was perhaps less suited to mathematical work or discussion, although among the libraries kept by some of the coffee houses there was the occasional copy of Euclid or a volume of mathematical recreations among the topical and political reading that made up their greater part.

Desaguliers' life seems to have been a busy, even a frantic one. His courses were advertised again and again in books and newspapers—the audience for public science was partly sustained by such advertising, through which readers were assured, sometimes in great detail, of the legitimacy and credibility, as well as the content, of the various lecturer courses which were available. Desaguliers' success was so great that rivals attempted to spoil his reputation by one means or another. He found himself obliged to deny in print that his 'apparatus' was 'worse than other Peoples or his Machines and Experiments fewer', and to deal with the unauthorized publication of notes from his lectures.

In the words of Jeffrey Wigelsworth, historian of science advertising, the world of popular lecturing was 'a public market where science and reputation were commodities to be advertised and sold.' Desaguliers' 'name recognition' was significant by this period, and he was repeatedly a victim of piracy. During 1720 he was also involved in an unpleasant dispute taking place in the advertisements of the *Post Boy* newspaper, when he and a pair of booksellers were involved in the production of rival English translations of Willem's Gravesande's *Introduction to Newtonian Natural Philosophy*.

Despite such problems, Desaguliers also found time to print a set of notes (in both English and French) for his lectures, and later, in 1734, to produce a published *Course of Experimental Philosophy*. He was an experimental demonstrator at the Royal Society, working with everything from prisms to steam engines, he worked as an engineer under private patronage, and he also did some work as a translator for the Society as well as translating books for publication. Desaguliers travelled to Cambridge, to Bath, and even to the Netherlands to lecture, but the centre of his activities remained London. He was admitted as an honorary member of the Spalding Gentleman's Society—at his own request—in 1734, proposed by John Grundy, but it is not clear that he ever attended a meeting of the Society.

His lectures and writings ranged across astronomy and natural philosophy: mechanics, hydrostatics, optics, and so on, with an emphasis on demonstrations and 'machines', including in later years an apparently very impressive planetarium. There was some degree of sensationalism about their presentation, with Desaguliers in sombre clerical black. A lady attendee described his house in Westminster as 'very much like the abode of a Wizard', filled with 'queer looking people called Philosophers'. Later she called the house an 'inchanted castle' inhabited by 'Conjurors'—Desaguliers would not be the last mathematical expert mistaken for a magician. Another audience member referred to Desaguliers' assistant as 'Whacum', the name of Sidrophel's assistant in *Hudibras*.

Despite any such concerns, Desaguliers was a massively successful popularizer of Newtonianism in the first half of the eighteenth century. An important part of his appeal was the reliable loyalty in political and religious matters he brought to his subject. The advertisement quoted above is taken from his 1728 publication *The Newtonian System of the World the best Model of Government*, a long poem setting out how George II's limited monarchy was like the Newtonian system of universal attraction, with a central sun whose power 'Directs but not Destroys' the liberty of the planets. Of course, the poem was also an opportunity for Desaguliers to display his powers as a popular scientific expositor: the book had very extensive notes and several quite complex diagrams elaborating upon Newton's mastery over defeated nature:

> Nature compell'd, his piercing Mind, obeys,
> And gladly shews him all her secret Ways;
> 'Gainst *Mathematicks* she has no Defence,
> And yields t'experimental Consequence.

As *The Newtonian System of the World* demonstrated, during this period Newtonianism was tending to become a universal philosophy of order and reason, incorporating the assumption that whatever your problems were, mathematics could help. In particular, perhaps, whatever thing you were interested in—odd-shaped barrel, fenland drainage project, constitutional monarchy, or solar system—mathematics could provide a geometrical or an algebraic construction that would serve as a model of it. Little of popular Newtonianism dwelt on the details of the mathematics—Desaguliers himself remarked that 'The Thoughts of being oblig'd to understand Mathematicks [had] frightened a great many from Newtonian Philosophy', and Newton's own mathematics was of course famously difficult to understand, even for specialists. But the propagation of Newtonianism by Desaguliers and others nevertheless changed the reputation of mathematics in eighteenth-century Britain, both among the ladies and gentlemen who attended his lectures, and beyond.

Theirs was a world of genteel interest and polite discussion—as at the Spalding Gentlemen's Society—in which natural philosophy might improve the mind just as, for others, geometry might:

> All hail PHILOSOPHY! Celestial Fair!
> Sent from above, replete with ev'ry Good,

> T'improve each striving Faculty within,
> To mend our Morals, and refine the Heart ...
> Which Way soe'er our darling Genius leads,
> We view the Footsteps of the DEITY!

But the Newtonian message ultimately promised the understanding and improvement of the whole world:

> Innumerable Uses flow to Man,
> Thro' ev'ry Branch of circulating Trade,
> From the just Lights which Nature's Plan affords.

Interest in these matters, as for the polite discussion of mathematics for its own sake, was by no means exclusively male, though evidence of female participation is quite scanty. In 1748 Thomas Rutherforth published a two-volume *System of Natural Philosophy*, based on lectures given at St John's College, Cambridge, and covering 'Mechanics, Optics, Hydrostatics, and Astronomy'. Nine women were subscribers, but that doesn't guarantee that they read the books. John Booth, lecturing in Dublin in 1745, offered a free ticket for one lady to every (presumably male) subscriber. The following year Henry Jones, poet and playwright, addressed those lady attendees in verse:

> 'Tis yours, with Reason's searching Eye to view
> Great Nature's Laws, and trace her winding Clue...
> To you, bright Nymphs, where Goodness charms us most,
> The Pride of Nature, and Creation's Boast,
> To you Philosophy enamour'd flies,
> And triumphs in the Plaudit of your Eyes.

Or, as Mary Scott put it in 1774 in her *Female Advocate or Women of the Future*, 'With matchless Newton now one soars on high, / Lost in the boundless wonders of the sky'. Some popularizers wrote books specifically for women, such as Francesco Algarotti's *Sir Isaac Newton's philosophy explain'd for the use of the ladies*, translated from the Italian in 1739 by Elizabeth Carter, whose thoughts on Desaguliers' lectures were quoted above.

Occasionally scientific and mathematical work was done by women. Caelia Beighton took over from her husband as editor of *The Ladies' Diary* for a period after his death in 1743. Mary Senex took over a map- and globe-making business from her husband after his death, and published an article in the *Philosophical Transactions* on his globes in 1749. In the 1780s and 1790s, and later, Caroline Herschel both assisted her brother William in his astronomical work and discovered a total of eight comets herself. She published her work in the *Philosophical Transactions* and in book form.

Another of the more prolific of women authors on mathematical natural philosophy was Margaret Bryan, who was active around the turn of the nineteenth century. We know little about her except what she reported or implied in her own books: she ran schools in at least three locations (Blackheath, London, and Margate), where she distinguished herself by teaching mathematics and other scientific subjects in some considerable detail to girls. The frontispiece of her books included a portrait of her with a classic selection of mathematical instruments, including a telescope and what seem to be an armillary sphere and a sextant. The picture also showed her two daughters: her first and, in her presentation of herself as 'Parent and Preceptress', most important pupils.

At first the parent and preceptress had a tendency to apologize for her boldness in publishing her courses of instruction. But over time the success of her books, together with the support of the mathematical professor Charles Hutton, led her to make fewer concessions to the 'false and vulgar prejudices of many, who suppose these subjects too sublime for female introspection'. Her three books ranged over astronomy, natural philosophy—including hydrostatics, optics, pneumatics, and acoustics—and geography, and they were all based on her own lectures. They remain lucid and readable syntheses of the state of those sciences in her day, and we may well imagine that the pupils to whom they were dedicated were grateful for the pains and the ability she had brought to her task.

She saw herself as a popularizer of the work of others, disclaiming originality and presenting her first book, *A Compendious System of Astronomy* (1797), as 'intelligible to those who have not studied the mathematics'. But its mathematical content was by no means insignificant: it contained some sophisticated trigonometry and spherical trigonometry, together with the use of the celestial globe. The first of her 'celestial problems', reminiscent of some of the questions in the *Ladies' Diary*, was 'To find the latitude of a place by the observation of two stars'.

Mathematics, in the hands of teachers such as Bryan, lecturers such as Desaguliers, or privately curious individuals such as the Spalding gentlemen, was in the Newtonian world a tool not just for understanding and predicting the world but also for describing and perhaps controlling and organizing it. This was a period, for example, when the discipline of 'political arithmetic' flourished. Its origin was found in the work of William Petty, an administrator in seventeenth-century Ireland, who attempted to use detailed statistical information to reform the organization of the state. Responsible for surveying the condition of Ireland, Petty had access to an enormous amount of data, including statistics on population and weekly death rates in both Dublin and London, which he supplemented with information from his own estates. The title page from a posthumous edition of his *Political Arithmetic*, from 1691, illustrates the scope of his ambitions to quantify political information.

> *Political Arithmetick*, or, a Discourse Concerning The Extent and Value of Lands, People, Buildings; Husbandry, Manufacture, Commerce, Fishery, Artizans, Seamen, Soldiers; Publick Revenues, Interest, Taxes, Superlucration, Registries, Banks; Valuation of Men, Increasing of Seamen, of Militia's, Harbours, Situation, Shipping, Power at Sea, &c. As the same relates to every Country in general, but more particularly to the Territories of His Majesty of Great Britain, and his Neighbours of Holland, Zealand, and France.

In the mid-eighteenth century the same impulse gave rise to such works as Arthur Young's *A Six Months Tour through the North of England, containing, An Account of the present State of Agriculture, Manufactures and Population, in several Counties of this Kingdom* (1770), or Joseph Massie's *Calculations of Taxes for a Family of each Rank, Degree or Class* (1756). The latter, as far as possible, tabulated the social structure and income distribution of England and Wales, producing an invaluable resource for the study of Georgian society.

At a time when the increasing complexity of public business was tending to push amateur gentlemen out of politics, political arithmetic provided an alternative way for some to contribute, or attempt to contribute, to the ordering of the nation. The genre also received, in 1776, no small impetus from the example of Adam Smith's *Wealth of Nations*, possibly the period's most influential book and an important example of the application of quantitative thinking to questions of national life and prosperity.

For some, political arithmetic and the data it gathered about Georgian society were supports for a conservative ethic of personal industry, work on the land under the wise rule of 'Farmer George' (George III, in the popular phrase), and suspicion of social change in the form of new wealth or of demographic shifts. For others it could assist the exploration of politically sensitive topics such as Britain's economic relations with its colonies or with Ireland, or it could promote political *ra*dicalism: mathematics, then, at the service of a proposed *re*ordering of the world. An example is provided by Richard Price, dissenting pastor, demographer, and radical. Price was taught mathematics at school in the 1730s by a friend and disciple of Newton, and married the daughter of a speculator ruined by the South Sea Bubble. In addition to his work as a pastor and his publications on moral philosophy he wrote on the subjects of poverty, demography, and the national debt. He contributed, for example, actuarial tables to a

1772 *Proposal for Establishing Life-Annuities in Parishes for the Benefit of the Industrious Poor,* and published an *Essay on the Population* in 1780 in which he (wrongly) identified a decline in the population of England and Wales.

Price's enthusiasm for the French Revolution and involvement with, among other groups, the so-called 'Revolution Society' in England made him a notorious and controversial figure, derided by Burke as 'the calculating divine'. Events in North America and in France, reflected in England by the radicalism of people such as Price, were changing the national mood to one of anxiety by the later eighteenth century, an anxiety which had among its focusses such quantitative matters as the handling of the national debt and the regulation of moneylenders. ('England's fate, / Like a clipped guinea, trembles in the scale!' wrote Sheridan in 1779).

It was in this context, aptly enough, that William Playfair (brother of the John Playfair well known for his translation of Euclid) brought forth the invention of graphs as a way to represent chronological series of numerical, and particularly economic, information. In his

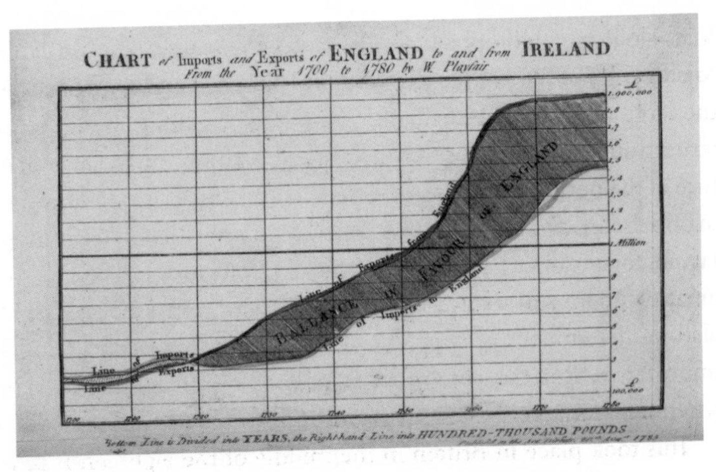

FIG 13 One of William Playfair's graphs, in his *Lineal Arithmetic*, 1798

By permission of the British Library (8503.ee.18)

Lineal Arithmetic of 1798 Playfair answered stiff criticism of his invention from those who evidently considered it an illegitimate application of mathematics.

> Suppose the money received by a man in trade were all in guineas, and that every evening he made a single pile of all the guineas received during the day. Each pile would represent a day, and its height would be proportioned to the receipts of that day, so that by this plain operation *time, proportion,* and *amount* would all be physically combined.
> Lineal arithmetic then, it may be averred, is nothing more than those piles of guineas represented on paper, and on a small scale, in which an inch (perhaps) represents the thickness of five millions of guineas, as in geography it does the breadth of a river, or any other extent of country...
> I have succeeded in proposing and putting in practice a new and useful mode of stating accounts... as much information may be *obtained in five minutes as would require whole days to imprint on the memory in a lasting manner by a table of figures.*

Thereafter, graphs became one more item in the mathematical toolkit for describing, understanding, and organizing the world.

Mathematics could organize time, too. Sundial time was available to all, and one part of the literature on practical mathematics was the large number of books and pamphlets teaching how to construct a sundial for almost any latitude or orientation. In the literature of the period sundials made of unlikely things (people's noses, in one almanac passage) and hour-glasses filled with unlikely things (wine, for instance) sometimes featured as forms of mock mathematics. As we will see in the next chapter, astronomical time could also be the subject of sophisticated computations with national importance. But the computation of time affected more people when time came to be *re*organized on mathematical grounds.

This took place in Britain in the middle of the eighteenth century, when the country replaced the Julian system it had employed hitherto with the Gregorian calendar that was by then in use across much of Europe. The need for change seemed quite urgent to some. By the early eighteenth century Britain's calendar differed from much of Europe—and from the seasons—by eleven days, with inevitable inconvenience when communication or trade had to cross the Channel. For years, many almanacs had carried a double column of English and continental—Julian and Gregorian—dates to help with the resulting confusion.

Isaac Newton had developed, but not published, proposals for a reform of the calendar: he would have done away with leap days in those years which were divisible by 100 but not by 500, and also added a second leap day in years divisible by 5,000. John Theophilus Desaguliers published with his *Newtonian System of the World* in 1728 a more orthodox proposal to adopt the Gregorian calendar. It took the form of a poem, called *Cambria's Complaint*—Wales was imagined looking over the *Oxford Almanac* and complaining about the disturbance caused to St David's day (1 March, which was also Queen Caroline's birthday) by the imposition of a leap-day during February. But there remained the problem of exactly how to change from the Julian calendar to the Gregorian. Desaguliers' proposal was an ingenious one:

> If his present MAJESTY reigns as long as Queen *Elizabeth*, that is 44 Years, and there be no Leap-Years in that Time, our Reckoning will advance 11 Days, and bring our Style to agree with the *Gregorian*.

In other words, the temporary abolition of leap days would, within a few decades, bring the two calendars into agreement. Desaguliers judged this preferable to the confusion that might result from the more obvious solution of having a year with eleven fewer days than normal.

> Had such a Chasm of Time with us been made,
> Bless'd as we are, with Opulence and Trade;
> How sorely had we felt our selves distress't
> In Landed and in Money'd Interest?

> Bills, Bonds and Leases wou'd have shifted Date,
> Affording Matter strange for Law-Debate;
> Tenants had murmur'd at Approach of Rent,
> And all our Stocks had varied Two *per Cent*.

Neither Newton's scheme nor Desaguliers' was adopted. In 1751 a Calendar Act was passed mandating the removal of eleven days from 1752 and the use of the Gregorian calendar in Britain thereafter. Almanacs took some pains to explain why 'September hath only XIX Days in this Year', setting out the fact that in the centuries since the start of its use the Julian calendar had moved gradually away from the seasons, because of a small difference between the average calendar year and the true solar year. They explained the new system, too, with ninety-seven leap years every four centuries, providing a calendar year which on average was a better approximation to the solar year, and which would therefore drift much more slowly relative to the seasons. The Stationers' Company continued until the 1780s to insert into some almanacs—particularly Poor Robin's—a one-page statement of how the change was to be applied—apparently as much as thirty years after the change they believed it could still be a source of confusion for some:

> By Virtue of an Act, made in 1751, for Alteration of the Style . . . the Use of the *Julian* Account, or Old Style, heretofore followed in this Country, ceased on the second of *September* of . . . 1752; and by dropping or leaving out eleven nominal Days, and calling the next, which would have been the third, the fourteenth, the New Style took Place.

The statement went on to explain that 'Fixed or Immoveable Feasts' were to be celebrated on the same 'nominal Days' as before (so Christmas, for example, would remain on 25 December), coming eleven days earlier than they would otherwise have done. On the other hand, 'all Things depending on them, such as the Opening or Inclosing of Common Fields and Pastures, the Holding of Fairs and Marts, the Payments of Rents and Annuities, and the

Commencement or Extinction of many private Rights and Matters of Property', which might be expected also to move by eleven real days, were directed in fact to be 'Kept, Observed, and Performed, on the same natural Days of the Year on which the said Feasts would have fallen, if this Act had not been made'. So, for example, the spring rent-day, 25 March (or 'Lady Day', the feast of the Annunciation), was shifted to 5 April, henceforth to be called 'Old Lady-Day'. (There it remains, which is why the tax year still ends on the fifth of April.) Some almanacs continued to print a double column of old-style and new-style dates to help with such shifts.

It must have been a troubling change to live through, and apparently it took a good while to get used to, particularly because it wasn't completely self-evident which annual dates should move, like rent day, and which should not, like Christmas. There were serious concerns that local fairs and markets would arrive too early for the produce that was supposed to be sold at them—growing vegetables didn't know that eleven days had been omitted from the year. Regulations linked to the calendar, such as those governing when herring boats could sail, were suddenly found to have shifted, illogically, relative to the seasons, with the result in some cases that part of the productive season would be lost unless the law was broken. The *Gardener's Kalendar* carefully set out new dates of planting and harvesting for crops, and many almanacs took to printing more detailed lists of the dates of local fairs in order to help with any confusion. City folk, whose calendar was rather easier to recalibrate, were not always sympathetic.

There were, similarly, concerns that the implications for taxation and contracts might be unfair. Some complained of deliberate bad faith: landlords, for instance, who allegedly began new leases on the new Michaelmas Day while letting old leases run until the old Michaelmas Day, charging rent twice for the eleven days in between. Others insisted on continuing to celebrate the major feasts on their 'real' dates. To the end of the nineteenth century there were communities that kept old-style Christmas: 'Auld Eel' in north-east Scotland, where it persisted as late as 1900. *Old Poor Robin* continued to print a double column of old-style and new-style dates, and to give the dates of a selection of old-style celebrations such as 'Old Candlemas' (14 February) and old Christmas day until the 1820s.

Again, real or feigned confusion could also be humorous. A correspondent of the *Gentleman's Magazine* in the month of the change, September 1752, wrote that he felt he would 'run mad' at the experience of the fourteenth of the month following immediately upon the second. 'Have I slept away eleven days in seven hours, or how is it?' He alleged that he had 'lost a wife' over the affair, an apparently somewhat waggish lady:

> I have solicited the most amiable of her sex these five weeks. She fixed the day for the tenth of September and gave a bond of 10,000 pounds for performance. That day has disappeared and my lawyer says my ten thousand are not worth a tuppence.

He blamed, of course, the mathematicians: 'A fine affair, that a man must be cheated out of his wife by a parcel of mackematicians and almanack makers before he has her'.

So although the disasters evoked by Desaguliers did not really take place there was, as well as laughter, perfectly reasonable confusion and well-founded resistance. There was a sense, too, that national pride had been damaged. Desaguliers had tried to reassure those of tender conscience that 'The *Roman* and *English* Style may certainly be reconcil'd, without any Danger of imbibing any *Popish* Doctrines in Religion'. But some were not so sure. The poet Elizabeth Tollet wrote that:

> Britain, 'tis true, was hard to overcome,
> Or by the arms, or by the arts, of Rome;
> Yet we allow thee ruler of the Sphere,
> And last of all resign thy Julian year.

Hogarth created a well-known slogan a few years later, in 1755, with his famous print of an 'Election Entertainment'. There, a broken placard read 'Give us our eleven days!', and generations of readers have perpetuated the belief that the depicted riot was a 'calendar riot'. More elaborate versions of the tale have the vulgar mob believing their lives were being shortened by eleven days. But recent scholars have found that such 'calendar riots' are no more than a myth. One historian, Robert Poole, writes that 'the riots, like the Snark, are universally known but defy detection'. He finds 'not a ripple' concerning actual riots in the historical sources of the period. Indeed, Hogarth's famous print—a piece of propaganda—seems to be the only contemporary 'evidence' for them. Another, Elizabeth Jane Wall Hinds, remarks that:

> The most interesting fact about the calendar riots... is that they probably never took place. Discord, yes; refusal to hold market fairs on New Style days, yes. But full-blown riots were more than anything else a myth.

A myth, that is, about lower-class ignorance of what the change of calendar meant. The poet Susanna Blamire made an aged Cumbrian woman remark, years after the change, that:

> 'E'er since that time the weather has grown cold'
> (For Jane forgets that she is now grown old).

After it had taken place, the issue of calendar reform became the object of one of the more eccentric applications of numerical expertise seen in eighteenth-century Britain. In a display of somewhat misdirected loyalty 'H.J.' proposed in 1753 a second change, to a new 'Georgian calendar'. This was to have thirteen months of twenty-eight days each, named for the apostles (Matthias substituting for Judas, and St Paul making up the total of 13). Christmas day would be added outside the monthly calendar and 'Britons-Day' would occur as a leap-day every fourth year 'except every 132nd year'. Despite the apparent absurdity of proposing a further change just after that of 1752, the unknown author seems to have been entirely serious, and printed a whole year's calendar to illustrate how his system would work and where the various feasts, legal terms and rent days would fall. He suggested that the new calendar should be adopted two years hence, in 1755, and continue 'for the space of 10219 years, *viz.* till A.D. 11974', when, according to his calculations, an additional leap-day would be required to keep the calendar exactly in line with the seasons.

The Pancronometer, in which the scheme was proposed, also contained a proposal to replace decimal arithmetic with base eight, and a discussion of the likely length of the year before the Flood. It does not seem to have received even the dignity of ridicule in print.

In 1765, Richard Dunthorne was appointed to a role as 'comparator' or 'comparer' in a new mathematical project based in London. Dunthorne was the son of a gardener in Huntingdonshire. He had attended his local grammar school and, thanks to his evident abilities, became a protégé of the master of Pembroke College, Cambridge, Roger Long. By 1765 he was in his mid-fifties: he had assisted Long in the construction of a planetarium, an eighteen-foot sphere in which several people could sit and watch a simulated view of the motions of the heavens, and had mathematical publications of his own to his name, including a book about the moon and papers in the *Philosophical Transactions*. Like many of the practical mathematicians we have met in this book, he had also worked as a surveyor: he was Superintendent of the Bedford Level Company and had done drainage work, lock building, and other fenland projects. He had built an observatory and devised mathematical techniques for reducing the effect of the refraction of light by the atmosphere on astronomical observations.

His work as 'comparator' was rather different from anything he—or anyone else—had done before. He was employed in this capacity by Nevil Maskelyne, the Astronomer Royal, as part of a small team of 'computers' or 'calculators'. (The word 'calculator' more normally meant either an accountant, or a book of mathematical tables: *The assistant calculator, or cotton spinner's guide; being a complete set of tables, of the greatest use in the cotton spinning business*, for example, which appeared in Manchester in 1799.)

What were they employed to calculate? Tables for use in determining the longitude, one of the eighteenth century's most significant uses of mathematics in reorganizing time and space.

In 1765, Maskelyne developed a plan to accomplish that celebrated feat by comparing the observed position of the moon with its position in astronomical tables. It would require the construction, years in advance, of lunar and other tables of greater accuracy than had hitherto been available, but that now seemed feasible. In 1755 the German astronomer Tobias Mayer, using theoretical work by Leonhard Euler, had produced new tables for predicting the position of the moon to the required accuracy.

Mayer's tables had been tested on the 1761 voyage to St Helena to observe the transit of Venus. Using them to calculate the longitude from observations worked, but it took about four hours each time. Maskelyne described the computational techniques involved in a publication of 1763, *The British Mariner's Guide*, but he recognized that for most mariners more was needed: not data from which to calculate, laboriously, the moon's position, but the position itself. The calculations that needed to be performed at sea to find one's longitude would then only take about thirty minutes. So what was needed was a table of lunar positions, calculated and published in advance.

Maskelyne, and various individuals who had tried out the calculations he described, were sufficiently convincing about the feasibility of this method that the Admiralty created a new computing office. As its head, Maskelyne distributed the calculation work for the lunar tables among his team of 'calculators', who worked at their homes around England. Each calculator was to carry out a complete set of lunar calculations for certain months of the tables.

The calculators were thus required to be highly skilled, and were given a lot of responsibility. They were employed rather like artisans in a cottage industry. (Systems more like calculating 'factories', in which individual responsibility was small and individual skill requirements minimal, would be adopted, elsewhere, in the nineteenth century.) Maskelyne provided the paper and ink, necessary tables, and 'computing plans': instructions were written on one side of a folded paper setting out the steps of the calculation; on the other side was a blank table ready to be filled in by the computer. Once it had been checked, this could be sent directly to a printer for typesetting.

Maskelyne's original division of labour involved two computers, each doing half the calculations for the moon's motion: one the positions at noon, the other those at midnight. The third person—the 'comparer'—collated the two sets of numbers and checked them for consistency. For the other tables that had to be produced, the two computers each did the same, complete set of calculations for the 1767 tables, and the comparer checked them for discrepancies and corrected any mistakes. The system proved highly effective, and it would remain substantially unchanged for forty years. Historian of computing Mary Croarken reports an incident in which

> Maskelyne's system for catching errors was so good that it could even catch out computers who cheated. Two computers, Joseph Keech and Reuben Robbins were caught copying each other's work instead of working independently. They were dismissed and asked to pay the comparer for the time it had taken for him to sort out the mess.

The tables were published in an annual volume under the title *The Nautical Almanac*. It didn't have much in common with the almanacs of Poor Robin and his colleagues, or even with the more mathematical *Ladies' Diary*, but 'almanac' was a natural word for an annual compilation of astronomical information. It would become a model for later almanacs whose contents consisted largely or entirely of data, facts, and lists. The main feature was the table of lunar distances, namely the distance of the centre of the moon from the sun and from various stars at intervals of three hours. Each month filled twelve pages, by contrast with the two pages per month of a conventional almanac. Supplementary tables were published separately, containing information which did not change every year—such as logarithms or the conversion of degrees to minutes—as well as tables embodying special techniques such as Dunthorne's for correcting the effects of atmospheric refraction.

The first generation of Maskelyne's calculators took longer than he had planned to learn how to carry out the complex set of calculations, and the work fell behind schedule. The calculators were not working full-time on the calculations: they had other jobs, and could not find extra time to make up for the loss. Maskelyne had to bring two more computers onto the 1767 tables, who were supposed to be at work on the following year's tables, but even so the first *Nautical Almanac* was finished much later than planned, and it appeared six days after the start of the year to which it referred, much too late for sailors undertaking long voyages.

Captain Cook was able to take the tables for 1768 and 1769 with him on his voyage of 1768 to 1771, and in his journal he complained of the fact that they were not being published further in advance. But the process sped up rapidly, and by the late 1770s the *Nautical Almanac* was being published three years in advance, and from 1783 five years in advance.

Once the almanac was established, the Board of Longitude also attempted to ensure that it was on sale at ports across Britain, Europe, and America, and that ships' commanders and masters were sufficiently trained in its use and in the use of the instruments (a quadrant or sextant) needed to make the necessary observations. Less confident navigators could be guided through the sequence of calculations required to use the *Nautical Almanac* using a printed form. The consequences for navigation were very real, and the

Board of Longitude would ultimately give Mayer's widow £3,000 in connection with the success of the scheme.

Maskelyne's *Nautical Almanac* was a piece of 'mathematics avoidance', enabling sailors to benefit from some very sophisticated mathematics by doing a comparatively limited set of calculations themselves. Its advantages as an approach to the longitude problem were obvious. In the late eighteenth century a copy of the *Nautical Almanac* cost about two shillings and sixpence, and a quadrant a few guineas. Enormously cheaper than a precision chronometer, these tools were distinctly less liable to be rendered useless by exposure to the elements. As a result, writes Nicholas Rodger, historian of the British Navy, 'the method of "lunar distances" was preferred'. Yet

> lunars were not a panacea. They could only be observed about twenty days in each lunar month, and of course all calculations are subject to error. Crossing the Atlantic in 1776, Home took a lunar which put them 300 miles west of their true longitude, they nearly ran on Nantucket Shoal when they thought they were off Long Island. Merchant ships were still crossing the Atlantic without chronometers, charts or sextants far into the nineteenth century.

So 'all but the most daring navigators still made their landfalls' by sailing directly east–west or west–east, obviating the need to know the longitude.

During his forty-six years in charge of the *Nautical Almanac*, Nevil Maskelyne hired thirty-five computers and comparers. At least a third lived in London, but many of the others were scattered across England, with a particular concentration in the South West. Most were practical mathematicians who mainly did other types of work, such as selling instruments, and relatively few had any formal mathematical or university training. David Alan Grier, historian of human computers, writes that

> From what we know of his [Maskelyne's] staff, all of them came from the second tier of astronomical talent. Most commonly, they had demonstrated some skill at astronomy but lacked the resources or the

connections to acquire one of the prestigious scientific appointments at Cambridge or Oxford.

They required some mathematical skills to do the work, but also the capacity to repeat the same sequence of computations many times without becoming sloppy. Writes Mary Croarken:

> The work they did was slow and painstaking. Their only calculating aids were logarithm tables, paper and pen and on dark winter evenings the work had to be done by candle light to fit in around other daytime employment.

The others recruited in the first year, alongside Richard Dunthorne, ranged in age from their mid-twenties to their mid-fifties; two were from lower-class families and one was a Jew, and thus excluded from a university career. Another long-serving computer excluded from the universities—because she was a woman—was Mary Edwards, whose clergyman husband John was also a computer until his untimely death in 1784. Thereafter Mary took over the work—she may well have been doing most of it already—and continued as a computer for many years.

Computers were paid on a 'piece work' basis for each month of the almanac that they completed. The rate was nearly £6 per month in the 1760s and rose to nearly £19 by the time of Maskelyne's death in 1811. For some, that amount of work would take no more than a month or so, but others took up to six times as long.

There were, inevitably, a few cases of serious incompetence, where a computer produced work that could not be used and that had to be repeated by someone else. Given a system which involved recruiting and supervising individuals who lived and worked at a distance, that was hardly surprising. More remarkable was that after the first few years there were no disasters of sufficient seriousness to disrupt the publication of the almanac several years in advance.

In achieving this, the role of comparer was crucial. That individual bore heavy responsibility for detecting errors by individual computers or collusion between two, and was also responsible for the critical task of checking proofs of the typeset tables against the manuscript originals. A diligent comparer would write to each computer (there were never fewer than four of them) several times a month to supervise their work. After Dunthorne, later comparers were schoolmasters, clergymen, and surveyors: one was Charles Hutton, a mathematical professor at the Woolwich Military Academy, whom we'll meet in the next chapter.

There was, indeed, some exchange of individuals between Maskelyne's team of computers, the assistants at the Royal Greenwich Observatory, and the naval and military academies. One of the first of Maskelyne's computers went on to be head mathematics master at the Royal Naval Academy at Portsmouth, another became master of the Royal Mathematical School at Christ's Hospital. Several assistants from the Observatory did some computing on the almanac after they left. One, John Crosley, was also president of the Spitalfields Mathematical Society.

There was also some relationship between the contents of the *Nautical Almanac* and the astronomical information in the Stationer's Company's almanacs. Several of these—following the criticisms of the 1770s—were now under the editorship of Charles Hutton, mentioned above. He increased the quantity and detail of the mathematical and astronomical information some of the almanacs contained, and in 1775 he engaged Henry Andrews, one of Maskelyne's computers, to perform the calculations for *Moore's Almanac*. The simplified information which appeared in other almanacs may well also have been derived from Andrews' work.

Even *Poor Robin* was affected. Under the editorship of Thomas Peat a section was removed which had appeared in the almanac for some time, giving directions for finding the time by the position of the Pleiades. Peat replaced this with a more elaborate chart giving, for the same purpose, the time when the major planets would be nearest the Moon. As a tool for finding the time on a clear night the usefulness of this might reasonably be doubted: as an exercise in astronomical prediction it was something of a virtuoso piece, and it must surely have been provided by someone connected with the *Nautical Almanac*, quite possibly Henry Andrews or Charles Hutton.

Thus, Nevil Maskelyne's lunar tables came to affect the mathematical information read by many thousands of people in their yearly almanacs, as well as in the specialized publications he produced for navigators. Even in Poor Robin's almanac their effect was felt, demonstrating the reach of this most ambitious—and useful—of calculating projects.

We have seen much—and we will see more—of the ways in which mathematics was useful in Georgian Britain. For some, the implication was clear: mathematics was *the* way to manipulate and to understand the world. If you were an admirer of modern science you could point to Galileo's dictum that the book of nature was written in the language of mathematics. If your favoured philosophy was Platonism, you could recall that that great philosopher believed that 'God geometrizes'.

Mathematics could still deceive, though (as it did the man cheated of his wife by the new calendar), and it could still obfuscate. Oliver Goldsmith's *Vicar of Wakefield*, written in 1761–2, featured a debate between a wise child and a mocking adult:

> 'I hope,' cried the 'Squire, 'you will not deny, that the two angles of a triangle are equal to two right ones.'—'Nothing can be plainer,' returned t'other... 'Very well,' cried the 'Squire, speaking very quick, 'the premises being thus settled, I proceed to observe, that the concatanation of self existences, proceeding in a reciprocal duplicate ratio, naturally produce a problematical dialogism, which in some measure proves that the essence of spirituality may be referred to the second predicable.'

And in Boswell's *Tour of the Hebrides*:

> Lord Powerscourt laid a wager, in France, that he would ride a great many miles in a certain short time. The French academicians set to work, and calculated that, from the resistance of the air, it was impossible. His lordship however performed it.

Equally, the later eighteenth century was the period of the so-called 'sentimental revolution': it was becoming common for individuals to be concerned about the possibly dehumanizing effects of what a writer of a later generation would call 'facts, facts, facts'. This kind of anxiety about mathematics was not just about the socially exclusionary nature of a subject too precise for polite conversation, nor yet necessarily about the dangers of reducing human beings to mere statistics, but instead about the risk that mathematics itself was inherently opposed to beauty and to what was fine and ennobling for the human spirit. 'Order the Beauty even of Beauty is', wrote Thomas Traherne in the mid-seventeenth century. Writing in 1753, William Hogarth could admit both straight lines and curves as possible expressions of the beautiful, but by the 1750s Edmund Burke, in his celebrated essay on the sublime and the beautiful, could write that

> surely beauty is no idea belonging to mensuration; nor has it any thing to do with calculation and geometry.... What proportion do we discover between the stalks and the leaves of flowers, or between the leaves and the pistils?

Indeed, changes in taste had already begun to make a dislike for geometrical regularity publicly visible: 'our gardens, if nothing else, declare, we begin to feel that mathematical ideas are not the true measures of beauty'.

Was mathematics the proper tool for organizing the world, or not? The debate continued, as we'll see in the next chapter.

Chapter 7

'A compleat Officer of Artillery'

GETTING IT RIGHT

Robert Sandham entered the Royal Military Academy at Woolwich as a cadet in August 1750. He wrote to his mother soon afterwards:

> I believe I need not inform you of the caution that is required in choosing an intimacy among a set of young fellows whose most honourable epithet is *wild*, the generality of them bear the worst of characters, being ever engaged in riots and drunken broils, in one of which a Lieutenant of the Train was lately wounded in the hand, and has lost the use of his little finger.

It was not all riot:

> Since I have been at the Academy, I have drawn a Cannon and a Mortar-bed by a scale, and begun a Landscape after Mezzotinto manner; the French master has been ill ever since I came, so I have not seen him.

A letter in November gave further indications of progress both social and intellectual:

> I received the ham and fowls safe on Monday night; Mr Muller gives you many thanks... I have written all *Mr Muller's Artillery*, which is forty octavo pages; I am now constructing the plates with Mr Simpson; I am in multiplication of fractions; Mr Mossiott approves of my

> drawing; as to the French Master, I have not seen him these two months, but I can read *Telemachus* with the help of a Dictionary.

Mr Muller was the mathematics master at Woolwich, the same man we met in Chapter 6 as a member and mathematical informant of the Spalding Gentlemen's Society.

Later in the month:

> You may be certain Mr Muller did not take the ham and fowls amiss, by his inviting me sometime ago to spend an evening with him, he made me a bowl of punch and made me very welcome.

The riots continued, we may be sure, but Robert assured his mother that

> I have not been in a tavern since I have been in Woolwich, except once, which was on the King's birthday, which I could not avoid without being thought particular, or perhaps a disaffected person.

The mathematics of war had been a familiar topic during the sixteenth and seventeenth centuries and, like the other mathematical subjects we have seen in this book, it increased in its range and its ambition during the eighteenth century, with, eventually, more and larger printed books produced for instruction, self-instruction, and practical use. Some parts of military mathematics were also of enduring interest to the leisured classes as a form of intellectual recreation. The design of fortifications, in particular, seems to have been a popular gentlemanly hobby, and the planning of geometrically elaborate and not necessarily very practical military structures for various situations was the subject of a wide range of publications, from practical handbooks to large expensive volumes that were roughly the early modern equivalent of coffee-table books.

But many, of course, would be required to put military mathematics into practice. Despite the relatively placid reputation of eighteenth-century Britain—the age of good taste and common sense, the age of a 'polite and commercial people'—the country was also repeatedly involved in conflicts, partly as a result of the continental interests of the Hanoverian dynasty. They included the wars of the Spanish (1701–14) and the Austrian Succession (1740–8), the so-called War of Jenkins' Ear (1739–48) and the Seven Years' War (1754–63), and conflicts in both North America and India during the 1750s. Robert Sandham would die at Fort St David's in the East Indies in May 1755.

Under George III, of course, Britain fought again in North America and, following the French revolution, in the French revolutionary wars and the wars of the Napolenic period. The need to train military officers had seldom been so urgent.

This was the context in which Robert Sandham entered the Royal Military Academy. There had been an attempt to set up a military academy in the early 1720s, with a mathematical master appointed at that time, but it seems to have failed and left little detailed trace of its activities. It was re-established in 1741, under George II, at Woolwich, on the south bank of the Thames near London. Somewhat further east than the Royal Naval College and Greenwich Royal Observatory, the area had long been in use as a dockyard and naval station.

The academy answered the continuing professionalization of armies and the need for increasingly detailed knowledge, particularly of the military sciences such as that of ballistics. The stated intention from the first was to instruct 'the people belonging to the Military Branch of the Ordnance in the several parts of Mathematics necessary to qualify them for the service of the Artillery and the business of the Engineers'. Thus the Military Academy, later the Royal Military Academy, provided important teaching jobs for a small number of practical mathematicians.

The founding warrant mentioned six hours of theoretical instruction to be provided three days a week, given by two masters. While the second master, on a smaller salary, was to take the students through 'the Science of Arithmetic, together with the principles of

Algebra and the Elements of Geometry', the first master was to teach a long list of subjects beginning with 'Trigonometry and the Elements of the Conick Sections', plus practical geometry and mechanics, mensuration, 'levelling', and fortification, moving on to 'the Manner of attacking and defending Places', and special topics including the 'Doctrine of Projectiles', the names, dimensions, and composition of various pieces of artillery equipment, 'and the several sorts of Fireworks'. 'In general he shall teach whatever is necessary to make a compleat Officer of Artillery, or a compleat Engineer'.

If that were not enough, the remaining three days of the week at Woolwich were to be taken up with the learning—by a larger cohort of students—of a wide range of practical matters ranging from the loading and firing of guns to the estimating of quantities of military stores. This practical course was to include, every other summer, a mock battle on a very grand scale, involving earthworks 'of the largest Dimensions the Ground will admit', with trenches, batteries, mines, and repeated 'blowing up' and 'making breach'.

Remarkably, non-military subjects were not neglected in what already seems to have been a very full plan of study. By the 1760s the academy was employing a French master and a 'master for Classics and Writing' as well as the two mathematical professors, an assistant, a drawing master, and fencing master. By the 1790s a dancing master had been added to the strength. During the nineteenth century lecturers in other subjects, ranging from history to geology, would be added.

Despite this gradual widening of the curriculum, the emphasis on mathematics remained a very important feature of the institution. Together with a similar emphasis at the Royal Naval College and the training of youths for the Navy at the Royal Mathematical School at Christ's hospital, it gives a real sense that mathematics was *the* subject needed for certain careers in the armed forces.

By the 1770s the mathematical part of the course was quite elaborately planned, and an entrance examination established in 1774 seems to have been intended to ensure that Woolwich would not have to teach very basic mathematics to cadets. Required for entry were a knowledge of the four rules of arithmetic and 'a competent knowledge of the rule of three'. In the lower academy at this time, the first class studied the elements of arithmetic, the second class applied them to practice, and the third class studied vulgar and decimal fractions and the extraction of square and cube roots. By the fourth class the students were introduced to algebra as far as the solution of quadratic equations. In the upper academy the cadets turned their attention to geometry, which here was no recreation or beautification for the mind but a practical tool: Euclid in the first class, practical geometry and surveying in the second, with such trigonometry as the study of fortification required. In the third class, conic sections and mechanics, in the fourth, ballistics.

By this time the division of labour between the two mathematical professors had changed, and they were called the professors of mathematics and of fortification and artillery. The former taught the more theoretical material from arithmetic to conic sections, 'as also geography and the use of the globes', while the latter took charge of practical geometry and mechanics, fortification, and the theory of artillery. If very basic arithmetic did have to be taught ('as far as the rule of three') it was done by the writing master.

Over time, the mathematical ramparts to be scaled at Woolwich were raised still higher; in 1792 there were additions to the curriculum including 'estimating of revetments, ramparts, ditches, batardeaux, powder magazines, turned and groined arches, etc'. The entrance examination likewise became more demanding (and the minimum age of admission increased). From 1813, entrants were required to know 'Vulgar and decimal fractions, duo-decimals or cross multiplication, involution, extraction of square roots, notation and the first four rules of algebra, definitions in plane geometry' as well as English and French grammar. For older candidates further algebra could be required 'except cubic equations', plus the first two books of Euclid.

The mathematical professors who oversaw all of this learning were, on the whole, men whose names are known as much from their other activities as for their work at Woolwich. From its founding in 1741 until the 1820s a total of thirty-five individuals were appointed at various times to the mathematics teaching staff of the Royal Military Academy (fewer than half seem to have used the title 'professor'). They included authors of textbooks, keepers of schools, private tutors, and editors of periodicals including the *Ladies' Diary*. Some were fellows of the Royal Society, or of foreign scientific academies.

Naturally, the need for officers tended to fluctuate with the beginnings and ends of Britain's various wars, as for instance it did during the invasion scare of 1797–9. At times, therefore, the masters seem to have been able to indulge in quite considerable leisure without disaster, if not without remark, although at others the need for officers became so urgent that the opposite was the case: there were periods when even examinations were abandoned. As well as lecturing, much of the professors' activity consisted of research, notably on the hot topic of ballistics. Some also wrote textbooks or developed careers in what might today be thought of as the popularization of their subjects.

John Muller, the first professor of fortification and artillery, whom Boswell would call 'the scholastic father of all the great engineers which this country employed for forty years', had, as we saw in Chapter 5, worked at the Ordnance Office in the Tower of London and had wide interests beyond the mathematics of war. The first professor of fortification and artillery was a Monsieur Landmann, who had taught those subjects at the École Royale Militaire in Paris (apparently no Englishman could be found who was so well qualified). The pattern would continue: the first six teachers of fortification and of geometrical drawing were all either French or German.

One of the Woolwich professors was Charles Hutton, a prolific author whom we have met on a couple of occasions earlier in this book, as an editor of the *Ladies' Diary* and a comparer for Maskelyne's *Nautical Almanac*. He was certainly the most prominent practical mathematician in Britain in his day, and the range and scope of his activities and writings says much about the importance of practical mathematical subjects by the late eighteenth century.

Hutton was born in Newcastle in 1737, the youngest son of a colliery labourer, and he worked briefly as a coal-cutter. According to his most recent biographer Niccolò Guicciardini, 'His first teacher was an old Scottish woman who taught in Percy Street, where the Huttons lived'. He was later instructed in mathematics, English, and Latin by a local clergyman whom, at the age of about nineteen, he later replaced as mathematics teacher at a local school. Hutton went on to work in Newcastle as a mathematics teacher, where he was apparently able to command high prices for his expertise: allegedly double the terms of previous teachers. He also ran courses for mathematics teachers, probably based on his own book *The Schoolmaster's Guide*. And he worked as a surveyor, and contributed to the *Ladies' Diary* and other periodicals.

At twenty-seven he wrote a treatise on *Arithmetic and Book Keeping*, for use in schools: the first of his many books. His *Treatise of Mensuration* had more than a thousand subscribers and contained the first known published work of the well-known engraver Thomas Bewick. According to Hutton's eulogist, later writers on that subject were reduced to making abridgements of his book, there being nothing new left to say. His *Arithmetic* was in its fifteenth edition by the time of his death in 1823, his mathematical tables their sixth.

Somewhat later, Hutton did some work for the Royal Society. He became a Fellow in 1774 and served as assistant to the secretaries in charge of foreign correspondence from 1779 to 1783. He must, therefore, have added to his mathematical knowledge some acquaintance with foreign languages, since the position involved, at least at first, translating foreign letters and making extracts from foreign books for the use of the Society.

But his association with the Royal Society ended very unfortunately. Hutton was forced by a series of redefinitions of the secretarial post to resign from it. Patrick O'Brian, as biographer of Joseph Banks, the then President of the Society, writes that 'on the face of it [Hutton] had been shabbily treated'. The incident became a pretext for attacks upon Banks by other mathematician fellows, and for a vote of no confidence in his presidency. He won the vote, but acrimonious attacks upon him continued, although the support for them dwindled in the end to just three fellows, of whom the ill-used Hutton remained one. The allegation has nonetheless tended to stick—unfairly—that Banks was hostile to mathematics or wished to preside over a society containing nothing but amateur natural historians. In fact, although he could be firm and even high-handed, he was neither unpopular, incompetent, nor ruinous for the Society, whose longest-serving president he went on to be.

Nevertheless, a number of fellows left the Royal Society over the affair, and Hutton failed in his attempts to be either reinstated as foreign secretary or elected as secretary. Perhaps more importantly, the incident provided fuel for a growing number of writings on the subject of the proper place of mathematics in national life, many of whose writers seem to have felt a sense of impending catastrophe when they surveyed the present situation of mathematical learning and teaching in Britain.

We saw in Chapter 3 that the provision at English schools of even basic mathematical instruction could be hit-and-miss: the universities in this period seem to have considered mathematics teaching to be at best a rather limited part of their function. The English university curricula, in origin, possessed a relationship with the ancient set of seven liberal arts, four of which were mathematical: arithmetic, geometry, astronomy, and music. It is not clear how far any of the four were really taught in the early modern period. When professorships of geometry and astronomy were first set up at Oxford in 1619 the statutes, though they clearly envisaged that much of the teaching and writing done by the professors would be at a rather high intellectual level, also made provision for the teaching of basic arithmetic for those students who needed it, to take place in the geometry professor's room. We know from Pepys' experience that some students nonetheless managed to come out of the English universities in the seventeenth century with very little arithmetical competence.

The situation was gradually changing: the number of mathematical professors in the English universities rose by the addition of the Lucasian and Plumian chairs. The founding of the Smith's prizes at Cambridge in 1768 (they were funded by South Sea Stock) perhaps did something for the prestige of the subject there. But there was probably still much truth in the accusations brought by Hutton and his supporters that mathematics was sadly neglected in the English universities. Oxford had no mathematical scholarships until 1831.

In the 1790s John Robison wrote in his article on physics in the *Encyclopaedia Britannica* that: 'A notion has of late gained ground, that a man may become a natural philosopher without mathematical knowledge'. Charles Hutton himself advocated, in some of his writings, the reform of British mathematics teaching along French lines. In 1803, the Reverend John Toplis published similar remarks 'On the Decline of Mathematical Studies, and the Sciences dependent upon them' in the *Philosophical Magazine*:

> It is a subject of wonder and regret to many, that this island, after having astonished Europe by the most glorious display of talents in mathematics and the sciences dependent upon them, should suddenly suffer its ardour to cool, and almost entirely to neglect those studies in which it infinitely excelled all other nations.

Like Hutton, he castigated what he perceived as a preference for natural history and a reliance on a national reputation gained in the increasingly remote days of Newton. He blamed, in part, mere fashion, and in part a lack of patronage—either individual or national—suitable to the devotion and financial loss which attended mathematical mastery and publication. In part, too, he believed that an outdated fascination with geometry was causing problems, 'confined in its application, feeble, tedious, and almost impracticable in its powers of discovery in natural philosophy'. He recommended instead that attention be given to 'analysis'—in other words, to the continental tradition of calculus and infinitesimals—and that the focus on the study of classical languages ('a course of education, which insults reason and sets common sense at defiance'), be overthrown. He insisted that mathematics and its allied sciences should be the main part of an education, repeating the old ideas of the use of mathematics in disciplining the mind and inculcating a proper sense of reason and accuracy.

Such concerns continued to be expressed. John Herschel, writing in his 1827 'Treatise on Sound' in the *Encyclopaedia Metropolitana*, wrote that

> Here, whole branches of continental discovery are unstudied, and indeed almost unknown, even by name.... In mathematics we have long since drawn the rein, and given over a hopeless race.

Herschel was associated with the founding of the Royal Astronomical Society in 1820 as, in part, an attempt to improve matters.

More optimistic and perhaps more measured was Charles Atmore Ogilvie, in a prize essay written at Oxford in 1817 when he was in his twenties. Ogilvie—who would become tutor and senior dean at Balliol, a canon of Christ Church, and Regius Professor of pastoral theology—proposed that mathematics should be combined with classical studies to form the most complete possible education, and reiterated some familiar claims about the benefit of mathematics to the mind. It helped to 'fix the attention' and 'increase the power of memory, by arranging the thoughts'. It exercised and thereby strengthened the mental faculties and the power of concentration upon a detailed argument, helping to avoid vague thinking and thus errors of judgement, and it helped to organize the thoughts and thereby increased the power of the memory. Finally, it tended to place the mind 'in that cool and dispassionate state, which is most favourable to deliberation and to judicious decision', helping to free it from undesirable external influences and prejudices. Thus it was a natural and essential part of a course of study by which 'man will best be trained to become happy in himself, and an instrument of diffusing happiness around him'.

If Ogilvie was disdainful of the direct usefulness of mathematics in the arts and sciences, which belonged merely to 'the education of the skilful engineer and mechanic', he acknowledged that an understanding of mathematics was vital in helping to understand and judge the claims of various arts and sciences. If he warned against an excessive attention to mathematics as tending to damage the ability to assess evidence in less clearly deductive situations, he retained a high sense of its value, in moderation, as a propaedeutic.

Possibly the culmination of all this came in 1830 with the *Reflections on the Decline of Science in England* by Charles Babbage, Lucasian Professor of Mathematics at Cambridge: a work of somewhat the same tenor, though principally concerned to attack the Royal Society and its members. More ominously—perhaps more tellingly—Mary Shelley would echo the same theme in *Frankenstein* (1818):

> If your wish is to become really a man of science and not merely a petty experimentalist, I should advise you to apply to every branch of natural philosophy, including mathematics.

Many of those who participated in this discussion were writing—like Hutton—from a sense of personal grievance. With hindsight it is difficult to feel confident that the language of national disaster they employed was really proportionate to the case, and they were to some extent blaming the universities for their political anxieties, particularly those raised by revolutionary France. In the event Britain was not found to be unequal to the challenges of the early nineteenth century, nor did its national prestige founder because of the details of its mathematical education system. Outside the universities there was a busy world of mathematical activity in eighteenth-century Britain.

Hutton, indeed, had been appointed as a professor of mathematics at Woolwich in 1773. Like his predecessors he had no university training, and he and nine competitors for the job were subjected to a 'public examination' to test their knowledge. The others were mathematical teachers and practitioners from broadly similar backgrounds to Hutton: none had publishing achievements that could rival his, although one, Hugh Brown, had published a translation and commentary on Euler's treatise on gunnery. The examinations, conducted by three fellows of the Royal Society and an engineer colonel, lasted several days, covering a range of problems as well as such matters as the best teaching texts for various branches of science. Hutton was the clear victor.

At Woolwich he taught the curriculum described above and participated in its development over time. His *Course of Mathematics*, published in 1798, became one route to success at the cadets' entrance examinations, when a knowledge of its first sixty-five theorems was established in 1813 as sufficient to gain entry to the Academy. The *Course* was frequently republished and translated into a number of languages including, eventually, Arabic. If the military world was much of Hutton's professional life, it also fell heavily upon his personal life: one of his two daughters died a prisoner of war in Guadeloupe in 1794.

He continued to be prolific as a writer on mathematical subjects ranging from the solution of cubic equations to the use of mathematical instruments. Often, what might seem highly theoretical works contained some practical material: for Hutton mathematics was a toolkit of universal application, of which no part was without its use. Thus, for instance, his geometrical work on the *Elements of Conic Sections* included problems concerned with emptying and filling ditches with water: matters of evident interest to those he taught at Woolwich.

From 1786, Hutton was also compiling almanacs for the Stationers' Company, at least ten of them: the *Ladies' Diary, Vox Stellarum* by 'Francis Moore', *Merlinus Liberatus* by 'John Partridge', *Speculum Anni* by 'Henry Season', *Atlas Ouranios* by 'Robert White', the Cambridge sheet almanac, and several more. He was paid well over £100 a year for doing so. As we have seen, he took the opportunity to increase the quantity of mathematical education and usefulness contained in some of these publications. He edited a multi-volume collection of the mathematical material from the *Ladies' Diary,* and an even larger abridgement of the first 135 years of the *Philosophical Transactions.*

Possibly the most prominent of all his works was the *Philosophical and Mathematical Dictionary* of 1795–6. It covered a vast range of subjects, and although it did not escape criticism for giving excessive attention to those subjects he personally found interesting—the entries oriented towards engineering seem to be more complete than those on other mathematical sciences such as mechanics, optics, and astronomy—it contained a great deal of valuable information, revealing Hutton's huge knowledge of both British and continental mathematical work, and of the history of mathematics.

By his retirement in 1807, Hutton had made what contemporaries called 'a handsome fortune' from his practice and writing, and it is hard not to feel that his heroic labours deserved it. As Patrick O'Brian writes in his biography of Joseph Banks, Hutton 'was an exceedingly hard-working man—a Newcastle coalminer's son who could write FRS and LLD after his name could hardly be anything else'.

On the battlefield, more than perhaps anywhere else, there was a need for mathematical knowledge to take the form of instruments: preferably instruments as straightforward to use—and as portable—as possible. Some military instruments were simply adaptations or unanticipated uses of those from other fields. A gunner who needed to know how distant a target was, for example, might use some of the same instruments that were employed in peacetime surveying, employing triangulation to determine distances which could not be measured directly.

More specific to military use were those instruments directly connected with gunnery, such as the calipers and gauges used to measure the diameters of guns and shot, and the sights and levels which were used to set a gun's range and direction. Figure 14 shows a 'gunner's perpendicular' from around 1790: it's about eleven centimetres tall, and was used to help find the line along which a cannon would fire. It consists, essentially, of a spirit level with a steel

FIG 14 A gunner's perpendicular, *c.* 1790

© Museum of the History of Science, Oxford

stylus at its centre, together with a brass frame so that the whole could sit astride the barrel of a cannon. In that position it could be adjusted until it was level, and then the stylus lowered in order to make a mark on the barrel. (The use of a spirit level for this was an innovation of the second half of the eighteenth century. Until 1757, instruments bought by the Office of Ordnance carried out the levelling using a plumb line.) That mark would be at the exact top of the barrel: two such marks at muzzle and breech, joined with a chalk line, would show the direction of fire.

This particular instrument has a shaped case of wood and pasteboard, covered with fish skin. It has the plainness of an instrument intended for practical use, although the fact that it lasted long enough to end up in a museum may indicate that the amount of use it actually saw was small.

The maker's signature is that of Adams of London: presumably George Adams junior, son and successor of the George Adams whose plane table we saw in Chapter 4. The family firm continued in business, and the younger George was a writer of mathematical and scientific essays and lectures in his own right.

Once you had determined the line of fire of your gun and pointed it towards the target, you still had to set its elevation, and for that you needed another instrument specific to this type of work, a 'gunner's quadrant'. The Adams firm supplied them in large numbers: they were broadly similar to the quadrants astronomers used for finding the elevation of stars and planets (we saw Tom of Bedlam holding one in Chapter 1), except that instead of sighting along it you would lie it down your gun.

At Woolwich the cadets learned to use these instruments, of course. John Muller's description of part of their training, from his *Treatise of Artillery* of 1757, gives a vivid sense of what it was like:

> In the spring of the year, as soon as the weather permits, the exercise of the great Guns begins, with an intention to show the gentleman cadets and private men, the manner of laying, loading and firing the Guns, at

> various distances from the but or mark; and as the line or direction is not marked upon the Guns, they have a small instrument, called a perpendicular to find the centre line or two points, one at the breech and the other at the muzzle, which are marked with chalk, and whereby the Piece is directed to the target; this being done, a quadrant is introduced into the mouth, in order to give it a proper elevation, which at first is guessed at, according to the distance the target is from the Piece. When the Piece has been fired it is sponged to clear it from any dust or sparks of fire that might remain in it, and loaded; then the centre line is found again, as before, and if the shot went too high or too low, the elevation is altered accordingly. This way of firing continues morning and evening for a month or six weeks, more or less, according as there are a greater or less number of recruits that have not seen any before.

As Muller's description made clear, trial and error was still usual in the mid-century as far as finding the range of a gun was concerned. But in the Newtonian world there was a strong impulse towards using mathematics to do better.

The science of ballistics was one subject in which Hutton's research supported his teaching at Woolwich. The motion of projectiles was naturally a complex study, since the considerations of air resistance and the effects of the wind meant that the relatively straightforward picture provided by Newton's laws of motion in a vacuum was of little use in providing calculations accurate enough for practical use. It was a subject that exemplified the potential of practical mathematics to make highly significant contributions to the Newtonian world.

In England the study of ballistics had been revolutionized by the work of Benjamin Robins, particularly in his *New Principles of Gunnery* of 1742. His career included mathematical publications in the *Philosophical Transactions*, controversy concerning the nature of limits and infinitesimals, and the ghostwriting of Lord George Anson's *Voyage Around the World*. His candidacy for a professorship at Woolwich was blocked by opponents on account of his political writings—instead he became Chief Engineer with the East India

Company, and Captain of the train of the Madras Artillery in India. He died aged 43 or 44, in 1751.

Robins did much by both experiment and theory to understand the motion of projectiles when air resistance diverted them from the pure parabola predicted by Newton's laws of motion in a vacuum. He was the first to realize how great the air resistance was on fast-moving objects, and therefore to understand its importance for finding the path of flight and the range of a projectile like a cannonball. One eighteenth-century writer said that Robins 'was in gunnery what the immortal Newton was in philosophy', another that his work had created a 'new science'.

In his book Robins dealt first with the explosive force of gunpowder, employing both direct experimental work on the force produced by explosions (their power to raise a column of water), and also experimental measurements of the speed of shot using his ballistic pendulum, essentially a pendulum whose (very heavy) bob would both take the impact of the projectile being studied and, by means of a stylus, trace on a curved track its resulting displacement. A knowledge of the behaviour of pendulums enabled the speed of the shot at impact to be deduced. This could then be correlated with, for instance, the size of the explosive charge that had been used, with the result that the dependence of muzzle speed on charge could for the first time be determined for guns of various different sizes and shapes and for shot of different sizes and materials. Naturally all of this information was valuable for military work.

Second, Robins considered the effect of air resistance. Although printed theoretical discussions of ballistics went back two centuries, he was dismissive of many of his predecessors and their reliance on theory in preference to experiment. Some persisted in the use of the parabola as a model of a projectile's path even when this was clearly incorrect, while others relied on an even older view in which a period of 'violent' straight motion shaded gradually into a 'natural', perhaps parabolic motion. The parabolic theory gave a range of about sixteen miles for a cannon ball which in fact travelled less than three.

Even the best available theory on this particular subject, including that presented by Newton in the *Principia mathematica*, was not adequate: for very rapid motion it predicted air resistance 'greatly short of what it really comes out by experiment'. Robins refused to soften the blow: 'the Theory of Resistance of the Air established by Sir Isaac Newton... is altogether Erroneous when applied to the swifter Motion of Musket or Canon shot'. Instead, Robins devised experiments to measure air resistance directly or indirectly, resulting in a new theory of air resistance at high velocities.

The experimental basis of Robins' work was constantly in evidence: never more so than when he remarked that 'shivers of lead' might emerge from between the ballistic pendulum and its iron backing, and that standing at its side was therefore not to be advised, or when he mentioned the tendency of the musket barrels of his time to burst, 'as I have found to my Cost'. There were, as he put it 'Dangers, to the braving of which in Philosophical Researches no Honour is annexed'.

Robins organized his book in 'propositions' and 'scholia', a manner which recalled Newton, surely self-consciously. He avoided algebra entirely, possibly hoping for a relatively wide readership, although in fact the details of his experiments and deductions, and the complexity of his mathematical analysis, would probably have been enough by themselves to defeat many potential readers. No less a mathematician than Euler translated Robin's book into German in 1745, adding still more detailed mathematical analysis. In 1753 Euler solved the equations for ballistic motion in air, and published some of the results in tabular form.

By 1750 John Muller, the first professor of artillery at Woolwich, had written his *Treatise of Artillery*, in which he used and attempted to improve upon Robins' results, although he seems to have lacked the personal experience of systematic experiments which Robins brought to the problem. His book was not published until 1757, but it circulated as a manuscript before then, as we saw in the case of Robert Sandham, who copied it out in full.

Charles Hutton also published various works on the same subject, including a paper in the *Philosophical Transactions* on 'The Force of Gunpowder and the velocity of Cannon Balls', deriving from experimental work done by Hutton himself using Robins' ballistic pendulum. The paper won Hutton the Copley Medal of the Royal Society. Hutton's theoretical work on gunnery included detailed consideration of the effect upon the elasticity of the air of the firing of explosive charges, and the effect of air resistance on the path of a projectile. His practical conclusions included principally a recommendation to reduce the amount of 'windage'—the difference between the size of the ball and the bore of the gun—making it possible to achieve a given shot speed with up to a third less gunpowder.

The 'ballistics revolution' thus displayed a remarkable interaction of theory, experiment, and technology. Both Euler and Robins produced tables showing the behaviour of projectiles. Robins also made range tables (published posthumously) of sufficient accuracy for practical use with a gunner's quadrant, so that by the late eighteenth century his theoretical results could in principle be used—through the medium of tables and instruments—in practice.

But practitioners were not quick to take an interest in the new results until they had been promoted by a generation or more of teachers, including Robins himself and those at Woolwich. John Muller, indeed, was at first hostile to the use of mathematical instruments for pointing guns, claiming in his treatise on military mathematics that it was better for cadets to learn to point pieces using their own senses. This was, indeed, one of those cases where mathematical training on the continent was quicker on the uptake than that in Britain, and with real consequences. The young Napoleon would study Euler's work on artillery.

In 1764 Daniel Fenning, perhaps best remembered as a grammarian and dictionary writer, prepared the first edition of his *Young Man's Book of Knowledge*. It was designed to help navigate the world of knowledge that would be of use to an up-and-coming youth in the 1760s. The *Book of Knowledge* covered theology, natural philosophy, geography, music, and the mathematical arts of geometry, astronomy, and navigation. The contents were hardly original, and in many of its details the *Book of Knowledge* followed its predecessors rather closely.

The book became a minor classic in its genre: after Fenning's death his publishers thought it worth their while to continue reissuing and updating the book, commissioning new illustrations and wholesale revisions of some sections by other authors. Fenning's ambitions for the book were modest enough. He dedicated the *Book of Knowledge* not to a nobleman but to the Lord Mayor of London. It carried recommendations from a short list of schoolmasters and clergymen otherwise unknown to history. It was published jointly by a London bookseller and one based in Salisbury in Wiltshire, since Fenning (presumably) intended it at least in part for the local market in his own part of the world.

Fenning was aiming at self-improving youths and their tutors. In 1792 one of those tutors bought a copy of *The Young Man's Companion*, probably in Salisbury. William Ross was a schoolmaster in Holwell Burrow, Dorset—his was a copy of the fourth edition, and he paid sixpence more than the cover price for the book, perhaps to meet the cost of having it bound. He kept the book long enough to write his name in it several times, but soon he passed it on to one of his students, Richard Shittler of Holwell in Somerset. This favoured individual added various notes to the book, writing again and again 'Richard Shittler his book', perhaps to make his claim clear to the other boys in an unruly classroom or, more likely, to his brothers, one of whom did in the end get possession of the book.

Richard learned to imitate his schoolmaster's handwriting quite closely, and we cannot always tell which of them made some of the notes in the book. Someone wrote some moralizing advice on the front flyleaf: 'Virtue is a Garment most comely and precious'. Richard himself wrote 'Richard Shittler his Book God give him grace that into book but not to look but understand'.

From time to time splashes of ink suggest Richard working through the book with pen in hand, making notes or copying sections into an exercise book. Once or twice he marked a page by folding it down. When he worked through the example on finding the position of the sun at a given time, he seems to have placed his sheet of notes in the book and closed it before they had dried, getting ink on the text. And the discussion of the lunar month and year became the occasion for a catastrophic spillage of what was probably ink. He practised his arithmetic, briefly, on the back of one of the musical examples.

Richard kept the book for a decade, and presumably learned something from his perusal of it. By 1802 his family had moved to another Dorset village, Haselbury Bryant. The book passed to his brother John in 1805, and later to another brother, William. It was probably this youngest owner who took the book on its greatest adventure.

In 1811 the book's owner went to sea, sailing in a merchant ship in a convoy which left Spithead in May 1811. Scarcely legible notes written into the book record this, but no more: on one reading they might indicate that the destination was in the Baltic, but a note of the latitude ('44 North') suggests instead a southern course. Very possibly William Shittler—if it was he—went on more than one journey. Whatever his destination was, the book went with him. The notes were made on the back of one of the illustrations in the mathematical section of the book, and it was probably that section which led the owner to make room for the book in his sea chest.

For the uses of mathematics could take it even to the far side of the world. In Chapter 6 we saw how mathematical calculation taking place across England was helping sailors to find the longitude from the 1760s onwards. Navigation at sea presented other difficulties, too, and they were addressed in a multitude of practical works.

The nature of navigation and the skills required of those who did it were significantly changing during the eighteenth century. The quantity of mathematical information and the number of tables in seamen's manuals was sharply increasing: after the introduction of the *Nautical Almanac* quite considerable computational competence began to be expected in those who would navigate over long distances.

Apparently simple questions—where I will end up if I start at such-and-such a position and sail on such-and-such a course?—are very much more difficult to answer when they are to be carried out on a sphere than on a flat surface, and indeed for such work the plane geometry of the first few books of Euclid or the plane trigonometry of basic mathematical textbooks was of only limited use. The elements of spherical trigonometry—dealing with triangles on the surface of a sphere—needed to be assimilated by those who wished to navigate over long distances by anything much better than guesswork.

Thus Fenning's book included a long section on trigonometry, setting out in a rigorous to-the-grindstone fashion the 'Seven Cases of Practical Trigonometry', the 'Six Oblique Cases' and the sixteen cases of spherical trigonometry. Each case presented the student with some combination of sides and angles in a (right-angled) triangle and asked for the remaining sides and angles to be found. Each case had its rule, to be learned by heart, and Fenning assumed that trigonometric quantities such as sines and cosines were being read out of a table which gave their logarithms, giving his calculations an unfamiliar appearance to the modern eye.

Thus Fenning's first plane case with right-angled triangles was this: 'The two acute Angles A and B, with the Base AC being given, to find the Perpendicular BC'. His rule was that as the tangent of the angle B is to the base AC, so is the sine of ninety degrees to the perpendicular BC: this he set out in a vertical column of working, employing logarithms to six or seven figures. Spherical trigonometry was perhaps conceptually harder, but computational strategies of the same general type were again set out to be learned and used. It must have been hard on the student who had just mastered the thirteen plane 'cases' to be presented with a further sixteen.

Later in the book all this was applied to navigation, first in a set of problems in 'plain Sailing', where the surface of the earth was assumed to be flat, and plane trigonometry was to be applied to answer questions such as this:

> Suppose a ship to sail SE by S . . . 124 Leagues; I demand her Difference of Latitude and Departure from the Meridian, *viz*. the Longitude she is then in.

Out came the rules for working plane triangles, and the problem was solved in a few lines (five degrees, nine minutes and eighteen seconds).

Then, at a more advanced level, the earth was considered a sphere and 'Circular Sailing' was considered, where spherical trigonometry had to be used and problems such as these were to be solved:

> Coming off the main Ocean, I had the Sight of a Cape, and intended to sail to it; I find it to bear from me NNW, and by Computation 33 Miles Distance. But having continued my Course N for 36 Miles from this Observation; I there anchored: Now I desire to know how the said Cape now bears, and its Distance from me.

The truly hardy were also given the opportunity to master more rapid but more conceptually difficult techniques involving either Gunter's compasses, or Mercator's chart, tools which enabled plane sailing to be improved upon without the explicit use of spherical trigonometry.

If William Shittler did take his copy of the *Companion* to sea in order to study this nautical material, he was quite probably hoping to better himself thereby—perhaps he was a midshipman hoping to pass for lieutenant. There seems to be no record that he ever did.

We don't know how his story ended, but somehow his book found its way back to England, and to a book dealer in Berkshire. Like some of the other books and objects we have seen, this copy of *The Young Man's Companion* provides a rare connection with the mathematics of ordinary lives in the eighteenth century, as well as showing something of the range and power which mathematics had now acquired.

Chapter 8

'The terrible *pons asinorum*'

PLAYING WITH IT

'Numbers are applicable even to such things, as seem to be govern'd by no rule, I mean such as depend on Chance'.

So wrote John Arbuthnot in his discussion of the usefulness of mathematics early in the eighteenth century. One of the boldest new manifestations of the power of mathematics was its developing ability to provide laws even for 'such things, as seem to be govern'd by no rule': to provide for the operation of reason even in situations where detailed causes could not be discerned for what was taking place.

The national lottery is one example of such a situation. The first English state lottery, the 'million adventure' was in 1694, and they continued throughout the eighteenth century. In a period when monarchs (Anne), princes (Wales under George III) and prime ministers (Fox) were sometimes notorious for their gambling, the lotteries offered the opportunity of a relatively modest flutter to a much larger class of people. Sanctioned by the state, it had a respectability that other forms of small-scale speculation lacked.

The state lottery was avowedly a scheme for raising money for the Government by a means more popular than taxation. (In fact it raised a loan: a ticket entitled its bearer to interest and, at a date chosen by the government, repayment. Prizes, too, took the form not of cash but of joint stock.) In some years the money was for a specific purpose such as, in 1736–7, building Westminster Bridge. In some years the government granted to individuals 'the power of disposing of their effects by way of chance': in other words in the unacknowledged hope of raising more money from them than they were really worth. Thus there was a 'museum' lottery in 1773–4, when the contents of a private museum were divided up and offered as a total of over 400 prizes, each winner to be entitled either to sell up or to participate as a proprietor of the museum.

The lotteries were popular from the first. The taste for risk to which they catered was quite a modest one—although buyers of lottery tickets were optimistically called 'adventurers'—and their potential to make new fortunes was distinctly limited compared with the stock market. Controversy about the ill effects of encouraging this form of gambling could be met by the argument that it was better than people betting the same money elsewhere: on the horses or on overseas lotteries.

The lotteries nonetheless attracted some adverse comment, and some ridicule. In 1730 a mock 'scheme' proposed to replace the marriage market with a lottery in which the prizes would be husbands, drawn from a group which would include 500 lawyers, 21,000 publishers, two lords and two 'scotchmen': each to be supplied with either a job or an income.

> *N.B.* If any young Ladies in a private Family should desire Tickets to be sent them...they must send a Certificate of their Age, Stature, Complexion, *&c.*

Some young ladies did indeed circumvent the restrictions of their circumstances in order to obtain tickets for the real lottery. In 1741 Elizabeth Teft contributed a poem, anonymously, to the *Gentleman's Magazine*, asking for generous volunteers to buy lottery tickets

FIG 15 Rowlandson, 'Cooper's Hall, Lottery Drawing', 1808
© Historical Picture Archive/Corbis

for her. A syndicate was formed to do so, and in mid-1742 the magazine printed a poem from Elizabeth announcing that she had not won.

Lottery drawings became major public spectacles, attended both by a considerable degree of disorder but also by a slightly ludicrous solemnity in the announcement of the winning numbers. One contemporary described the scene: 'a number of drunken ragged blaspheming wretches' on a fool's errand to see whether they had won, 'cursing themselves for their folly'. One pastime during the time of the drawing was to bet on whether the next number drawn would be a prize or not.

By the late eighteenth century the numbers were large. The state lottery of 1799, drawn the following year, made available 55,000 tickets and promised a total of £500,000 in prizes, ranging from two top prizes of £30,000 down to over 16,000 prizes of just £18. Nearly one ticket in three would receive a prize. Tickets were correspondingly expensive, and their retail price rose as the drawing neared: by December they were on sale for nearly £16.

By the later part of the century lottery tickets were being handled by large brokers: Johnson and Co., for example, had four separate offices in London to handle their trade. There they sold such sophisticated deals as their 'twelve shilling adventure', which offered a 5 per cent share in four different lottery tickets.

For those who wished for something cheaper, canny dealers also sold 'insurance' on lottery numbers. The purchaser of insurance on a given number would receive a payment in the event that that number was drawn as a 'blank' or, alternatively, in the event that it was drawn as a prize. The payments themselves could take the form of lottery tickets: as many as 500 a day were reportedly distributed by this means on some occasions. In principle, insurance sold for a couple of shillings might win you a lottery ticket which then itself won a large prize: a case was reported in which the insurers 'Messieurs Shee and Johnson, of Exchange-Alley' handed out a ticket which won £10,000 the next day.

Less sophisticated schemes used small shares in lottery tickets to inflate the value of trinkets: buy an over-priced handkerchief and get a free share in a lottery ticket. Or, in one enterprising London eatery (Fuller's Eating-house, in Wych-street: the scheme was advertised in the *Morning Chronicle*), buy threepence-worth of meat for sixpence and get a free chance in the lottery. Others resorted to plain fraud, in which the tickets were never owned by the vendor in the first place.

By the later part of the century the 'Frauds and Tricks' attending the lottery were a matter of regular comment in print, and certain celebrated cases such as that of 'Charles Price' alias 'Patch', a dishonest lottery-office keeper and forger who used the purchase of lottery tickets as a way to launder fake bank notes, had become bywords, as had the law's powerlessness to do anything about most such cheats.

> I have seen the Office Keeper's windows and shops demolished, by a deluded and justly enraged mob, who have been ruined by the purchase of tickets, snares, chances, and insurances thereon.

Cautionary tales circulated, such as that of 'John Doyle' who lost his quarter's pay on the lottery, pawned the silver and lost the proceeds as well, and finally turned highwayman in the hope of gaining money for more tickets (he was hanged for it). Henry Fielding penned a farce about lottery adventurers, in which inevitably the fortune was proved imaginary and the gold-digging husband disappointed:

> A Lottery is a taxation,
> Upon all the fools in creation;
> And heav'n be prais'd,
> It is easily rais'd,
> Credulity's always in fashion:
> For folly's a fund,
> Will never lose ground
> While fools are so rife in the nation.

Tickets, shares, insurance, and the rest all exploited buyers' inability to estimate what a given deal was really worth, and indeed their uncertainty about what the lottery tickets themselves were worth. During the drawing of the winning tickets—which could go on for weeks—speculation would reach a frantic rate, and the price of some deals would rise, not always rationally. Of course, a lottery is by its nature a bad gamble, an 'adventure' in which the most likely outcome is loss, not gain. Buying a lottery ticket is always, in one sense, irrational.

Unless, that is, you have a system.

John Molesworth was probably the most celebrated peddler of lottery systems in Georgian Britain. The popular success of his schemes—in the sense that people bought his books and pamphlets in large numbers—reached such a height in the 1770s that he was overwhelmed by correspondence from interested parties. He resorted to numbering the copies of his books and refusing to answer enquiries from those who couldn't quote the number of their copy.

He was—or claimed to be—quite a learned man, educated, in his own account, at Peterhouse, Cambridge, and the Inner Temple in London. Any knowledge of the law he had acquired at the Inner Temple seems to have been employed largely in staying just the right side of it, yet his university education enabled him to present with some show of conviction, not vague hints and opinions, but 'proofs' of his claims about the lottery and how to profit from it. As a demonstration of his good faith and altruistic intentions, and of his abilities as a calculator of probabilities, he published tables showing how much certain lottery insurance schemes were really worth to the purchaser. He lamented the fact of living in an age which gave relatively little value to new discoveries in the sciences, and which as a result had failed to reward him as his merits deserved.

> The man of liberal education and distinguished abilities, whose life, perhaps, has been devoted to intense study and useful researches, when he at last has succeeded in a favourite pursuit, and ventured to usher his performance to the world, is generally treated with contempt, and either shunned and despised as an impostor, or deemed a projecting mad-man; and so universally does this reward attend the authors of new inventions; that it amounts almost to a certainty, that a man of genius *must be poor*.

This man of genius proposed, essentially, that the mechanical apparatus—a wheel—used at lottery drawings did its work of randomization somewhat imperfectly. He claimed that it was therefore possible, at least in some circumstances, to identify a group of numbers which were more likely to be drawn for prizes than others: indeed, to identify numbers which were more likely to win specific prizes, because of the predictable sequence of operations that was used. Thus, in 1770:

> I found 25,000 tickets, half the number in that Lottery, would produce more prizes than the other half of the Lottery, in at the least, the proportion of 7 to 6, and that the numbers in my list were consequently worth more considerably than the others.

Molesworth claimed to have spent more than £2,000 'in procuring books and most curious wheels' upon which to experiment, and throughout his public statements he attempted to throw the emphasis onto verifiable facts, diligent experiments, and laborious calculations: the tools of confidence and credibility in the Newtonian world. He arranged for his list of the more fortunate numbers to be sealed before the 1770 drawing, announcing the fact in the newspapers. On opening, the list turned out to contain—he said—fifty-two out of the seventy-two numbers which won the top prizes, including all of the top five: the Lord Mayor of London, whose seal had been used, gave a certificate of the fact.

By these means Molesworth was able to persuade a number of individuals who looked into the matter in detail that there was something in his scheme: that his list of numbers did indeed contain more of the prize-winners than chance could account for. A random ticket had about one chance in 700 of winning a prize: his chosen tickets seemed to have slightly better than one chance in 500. This wasn't enough to make the lottery itself profitable for the astute purchaser:

> let it be considered, that the chances of Lotteries are seldom less than thirty per cent. against the adventurer, and I have never said that my tickets were above ten per cent. more valuable than others.

But it was enough to make the trade in these supposedly more valuable tickets profitable to Molesworth and his agents.

By 1774 he was promoting a scheme in which, by judicious investment in insurance on a small set of more fortunate numbers, 'two out of three adventurers' would be gainers. He asked no gratuity 'unless the drawing of the Lottery fully confirms my assertions; nor even then, from any but those who are fortunate'. The following year his promises were more ambitious, and included reimbursement for followers of his instructions who failed to win a certain proportion of prizes.

While his earlier claims had the ring at least of conviction if not of truth, the terms of play in these later proposals were complex enough to confuse even some mathematically astute commentators at the time, and to create lasting doubt as to whether Molesworth still credited his calculations or was now merely attempting to cash in on popular belief in their efficacy and his 'luck'. At a distance of more than 200 years it's hard to be sure how far Molesworth himself had ever believed in his claims, or whether 'the Molesworth plan' had ever been anything more than a scheme for inflating the value of certain lottery tickets for personal gain.

Contemporaries were divided about Molesworth's claims, partly because he was widely misunderstood—despite what seem to have been his best efforts to avoid such a confusion—as claiming he could identify the exact tickets which would win prizes, a mistake which naturally led to some sharp disappointments. Certainly his claim, repeated and expanded over the years, that certain lottery tickets were worth more than others was quite inflammatory from the point of view of those who bought, sold, or traded in them, and it impugned the frequent claim of those who ran and promoted the lottery to act 'fairly'.

Molesworth faded from view after a few years: it's not clear how much he had gained from his lottery scheme, or what he did with his 'genius' later in life. He may have been exceptional in his notoriety, but his scheme typifies the culture of 'luck' that surrounded the lottery and which led even the more respectable brokers to print lists of the prizes recently won by 'their' tickets. Firms with names like 'Goodluck' and 'The Lucky Lottery Office!' offered, unlike Molesworth, no explanation for their special relationship with Fortuna.

Review the Scheme, say I,
  Of wealth its *Advertiser,*
Your lucky *Star* then try,
  And BISH's Tickets buy, Sir.

> Each *Journal* tells his name,
> To Luck he is conductor,
> He'll *Pilot* you to Fame,
> Take him as your *Instructor*.

It is paradoxical that this culture of luck should have arisen at roughly the same time as the development of the mathematical theory of probability. Like the stock market, the lottery threatened to bring 'calculation' into disrepute. Mathematical practitioners responded, placing themselves on the side of the deluded purchaser and attempting to make a little profit from the lotteries in their own way by the publication of pamphlets on the 'doctrine of chances' and the real value of tickets, 'insurance', and other deals.

Samuel Clark was one, active in the 1770s. One of his pamphlets was specially printed for distribution during the drawing of the lottery: he signed each copy himself, apparently to guarantee its genuineness and to emphasize that he was placing his personal credibility behind his calculations.

He set out a brief exposition of the laws of chance and the idea of expected gain and loss, and showed some of the ways in which intuition about games of chance could be shown by calculation to be faulty, repeatedly quoting Abraham de Moivre's work on probability. To refute Molesworth's by now detailed and complex claims he employed a quantity of algebra probably beyond most of his readers, claiming that Molesworth and his team were the only gainers from the scheme. He characterized Molesworth as no better than a quack prognosticator, and belief in luck as a form of 'superstition' no better than astrology:

> The most eminent Writers upon the Laws of Chance utterly deny the Existence of such a Principle, and Men of the most refined Understandings do, at this Time, universally explode it....
> Let us see what the occult Sciences, by some called Astrology, or in the more vulgar Acceptation of the Word, Conjuration, will do for us, in

order to account for this surprizing Phenomenon of intuitive Knowledge in future Events.

A brazen lion placed in the middle of a reservoir throws out water from its mouth, its eyes and its right foot. When the water flows from its mouth alone it fills the reservoir in 6 hours; from the right eye it fills it in 2 days; from the left eye in 3, and from the foot in 4. In what time will the bason be filled by the water flowing from all these apertures at once?

Numerical fun didn't end with the lottery. Volumes of 'mathematical recreations' had been appearing since the seventeenth century. Thanks in part to the efforts of Charles Hutton, whom we met in the last chapter, they underwent something of a revival around the end of the eighteenth century.

The problem about the bronze lion appeared in his *Recreations in Mathematics and Natural Philosophy* of 1803. There were four volumes of the *Recreations*, and they aimed to teach the reader a good deal about mathematics and its applications under the guise of entertainment and idle curiosity. The work began with a short course in basic arithmetic, and a reader who followed it through to its conclusion would have learned some quite sophisticated algebra and geometry by the end.

Hutton's title page listed an enormous range of mathematical and philosophical sciences to be covered, including the modern innovations of magnetism and electricity as well as classic studies such as arithmetic and geometry and classic practices such as navigation and the making of sundials. There was, for instance, a section on political arithmetic, discussing such matters as the ratio of males to females in a given population, and the use of tables of life expectancy to answer such questions as this:

A young man 20 years of age borrows 1000l. to be paid, capital and interest, when he attains to the age of 25; but in case he dies before that period, the debt to become extinct. What sum ought he to engage to pay, on attaining to the age of 25?

Under 'arithmetic' alone the problems and puzzles ranged across what would now be called number theory and the properties of arithmetic, geometric and harmonic progressions, combinatorics and probability, and magic squares. In part, this encyclopaedic scope reflected Hutton's own interests, and his position as the compiler of the *Philosophical Transactions*' abridgement, editor of the *Ladies' Diary*, and author of his own mathematical dictionary. Few men in Georgian Britain—or indeed in any period—would have been better placed to bring together such a wide range of amusing, entertaining, yet instructional mathematics.

In its relatively serious-minded approach, Hutton's *Recreations* contrasted with the somewhat earlier, and quite popular, *Rational Recreations* of William Hooper. This was also in four volumes, and had first appeared in 1774. Hooper promised that 'the Principles of Numbers and Natural Philosophy are clearly and copiously elucidated, by a series of easy, entertaining, interesting experiments'. The emphasis was on 'pleasure and surprize' and it was meant for readers who liked things to be easy: 'The principles of each science are... laid down in a few plain aphorisms, such as require no previous knowledge, and very little capacity or attention to comprehend'. Hooper envisaged 'Philosophy, with his sober garb and solemn aspect, when led by the hand of the sportive nymph Imagination, decked in all the glowing ever-varying colours of the skies, may gain admittance to the parties of the gay and careless', where it would provide new pleasures for even the most jaded of appetites.

Hooper's was a thorough compilation. Even though the bulk of the work was taken up with the more spectacular natural sciences, the mathematical sections contained number tricks and card games galore:

> *A person privately fixing on any number, to tell him that number.*
> After the person has fixed on a number, bid him double it and add 4 to that sum, then multiply the whole by 5; to the product let him add 12, and multiply the amount by 10. From the sum of the whole let him

> deduct 320, and tell you the remainder, from which, if you cut off the two last figures, the number that remains will be that he fixed on.

Hooper had an eye for the memorable use of titbits of mathematics: his section on combinations included questions such as 'how many different sounds may be produced by striking on a harpsichord two or more of the seven natural notes at the same time' and 'the number of changes that may be rung on twelve bells'.

Hutton had found the bronze-lion problem not in Hooper but in a different predecessor: Jean-Étienne Montucla's 1778 volume of *Récréations mathématiques*. Montucla, coming from the milieu of the French 'Encyclopédists', had wide interests in mathematics and its history. His volume was an expanded and revised version of an earlier one from 1694, also in French, by Jacques Ozanam. That slightly chaotic collection contained straightforward number games, as well as some far from trivial pieces of geometry, optics, mechanics, and much more. The 'problems' ranged from geometrical constructions and guess-my-number games, to building sundials and weighing devices, and even shooting a pistol behind one's back (using a mirror).

Although Hutton would pronounce it 'rational and complete', and despite its popularity, Montucla judged Ozanam's collection of recreations to be 'very faulty, and incomplete'. He considered that a revised version was necessary both for that reason and because of the changes that had taken place in natural philosophy since his day.

Jacques Ozanam was by no means the beginning of this particular road. He undertook his volume of mathematical recreations as a response to what he believed to be the inadequacies of yet another predecessor, some of whose material he used, namely Claude Mydorge. Mydorge himself was revising the work of 'Henrik van Etten' (Jean Leurechon), whose 1627 *Récréations mathématiques* stands at the head of this particular genealogy. Hutton was impressed with neither of these ancestors: he called Leurechon's book a 'mere wretched rhapsody' and Mydorge's

> a confused collection of questions, the greater part of which are silly and childish, and expressed in barbarous language, sufficient to disgust any person of only common taste.

Etten's version of the brazen lion problem was, nevertheless, only slightly different from Hutton's 170 years later. He gave the information in a different order and placed the words in the lion's own mouth: 'Out of my right *eye* if I let *water* passe, I can fill the *Cisterne* in 2. *days*', and so on.

The development that had taken place by Hutton's day was in the approach to a solution. Van Etten took, to put it mildly, a minimal approach. He merely stated that 'the solution is by the Rule of 3. by a generall Rule, or by *Algeber*', leaving the reader potentially at a loss to understand how any of those practices was to be applied to this particular case. Subsequent generations of authors of mathematical recreations would differ in their view on how much help towards a solution the reader deserved, but few would be quite so unhelpful as this.

Hutton, by contrast, took a systematic approach, showing first that 'as the lion, when it throws the water from its mouth, fills the bason in 6 hours, it can fill $^{1}/_{6}$ of it in an hour', and so on for the other apertures. Then,

> By throwing the water from all these apertures at once, it furnishes in an hour $^{1}/_{6}$ + $^{1}/_{48}$ + $^{1}/_{72}$ + $^{1}/_{96}$, and these fractions added together are equal to $^{61}/_{288}$. We must therefore make the following proportion: If $^{61}/_{288}$ are filled in one hour or 60 minutes, how many minutes will the whole bason, or $^{288}/_{288}$, require: Or, as 61 is to 288, so is 1 hour, to the answer, which will be 4h. 43m. 16 $^{44}/_{61}$ seconds.

The lineage of this problem, as van Etten acknowledged, stretched back still further: it was by no means his own invention. It was taken, in fact, from the 'Greek anthology', a collection of ancient Greek poems and epigrams compiled—in the form that would have been known to van Etten—in the fourteenth century on the basis of several earlier collections. Among much material of other kinds the anthology contained problems and riddles, many of which were arithmetical in character, including several others which made (and make) regular appearances in the mathematical puzzle literature. In the anthology the bronze lion question was stated thus, according to a modern English translation:

> I am a brazen lion; my spouts are my two eyes, my mouth, and the flat of my right foot. My right eye fills a jar in two days, my left eye in three, and my foot in four. My mouth is capable of filling it in six hours; tell me how long all four together will take to fill it.

So with this bit of mathematical fun Hutton was engaging, in a way, with the 2,000 year history of Greek literature and its reception and transformation in Europe, as well as with close to 200 years of the publication of mathematical recreations in English and French. The line would continue, of course, through the nineteenth and twentieth centuries, and mathematics would continue to be a source of fun for generations more readers of the brazen lion problem and others like it.

> As Miller's Hussars marched up to the wards,
> With their captain in person before 'em;
> It happened one day that they met on their way,
> With the dangerous *Pons Asinorum*!
> Now see the bold band, each a sword in his hand,
> And his Euclid for target before him;
> Not a soul of them all could the dangers appall
> Of the hazardous *Pons Asinorum*!
> While the streamers wide flew, and the loud trumpets blew,
> And the drum beat responsive before 'em;
> Then Miller their chief thus harangued them im brief,
> 'Bout the dangerous *Pons Asinorum*!

"My soldiers," said he, "though dangers there be,
Yet behave with a proper decorum;
Dismiss ev'ry fear, and with boldness draw near
To the dangerous *Pons Asinorum*!"
Now it chanced in the van stood a comical man,
Who, as Miller strode bravely before him,
To his sorrow soon found that his brains were wheeled round,
As he marched to the *Pons Asinorum*!
O sorrowful wight, how sad was his plight,
When he looked at the *Pons Asinorum*!
Soon the fright took his heels, like a drunkard he reels,
And his head flew like thunder before him.
So rude was the jump, as the mortal fell plump,
That not Miller himself could restore him;
So his comrades were left, of 'Plumbano' bereft,
O pitiful plight to deplore him!

So wrote the 13-year-old poet Thomas Campbell 'in Mr. J. Miller's mathematical class' at Glasgow University in 1791. The *pons asinorum* is the fifth proposition of the first book of Euclid's elements, the bridge over which asses, so it is said, cannot cross (it also comes with a diagram that vaguely resembles a bridge): the two angles opposite the equal sides in an isosceles triangle are equal. The poem was apparently occasioned by the confusion of an over-confident fellow student in the face of the proposition, on examination day. A biographical sketch of the teacher, Miller (or Millar), provided by the university states that he struggled 'to improve teaching, in which he was said to have a genuine interest although he was remembered as being "subject to mental distraction and had little control over his students"'.

By the end of the eighteenth century, mathematics had become something to laugh about (again). A romantic sensibility perhaps tended to find less occasion to use mathematics as a metaphor—certainly in any positive sense—than the preceding two or three generations had done, but the early nineteenth century did have some sense of fun about the subject.

Some authors took a rueful tone about the universalizing art at which—alas—they were no good. A young man of the generation after Campbell eulogized mathematics but regretted his inabilities at it:

> They say you lead to grand results, and Science
>   Makes you unto her heaven a Jacob's ladder;
> So clouded though, we cannot see the sky hence,
>   And black-gowned students are a vision sadder,
> Nor promise half so much for what they spy hence,
>   As did the white-robed angels Jacob had a
> Glimpse of; but be that as it may, you lead to
> Things greater far than I can e'er give heed to.

This was Alfred Domett, who would go on to a varied career as a New Zealand politician and administrator, including a period as premier in 1862–3. He wrote his 'Ode to the Mathematics' in England at the age of nineteen. He blamed his 'scape-grace, wandering, weak, wool-gathering mind' for his poor relationship with mathematics:

> You teach mankind all, all that can ennoble 'em;
> Meantime I'm staggered by this plaguy problem!

But in the same breath he suggested the subject itself might well be to blame:

> Ye intellectual catacombs—where drones
> Of many an age have piled up musty bones!

Yet another young man produced around the same time the still more uproarious spoof 'The loves of the triangles', a skit on Erasmus Darwin's 'Loves of the Plants' from 1798. Like many another, 29-year-old John Hookham Frere (diplomat, poet, and translator) picked up on the incomprehensibility of mathematical jargon to outsiders, in a charming fantasy in which geometrical shapes were:

> Fair sylphish forms—who, tall, erect, and slim,
> Dart the keen glance, and stretch the length of limb;

To viewless harpings weave the meanless dance,
Wave the gay wreath, and titter as they prance.

He also drew satirical attention to the continuing gulf between practical and gentlemanly mathematics—Euclid for gentlemen but arithmetic and quasi-algebra for tradesmen:

Stay your rude steps, or e'er your feet invade
The Muses' haunts, ye sons of War and Trade!...
Debased, corrupted, grovelling, and confined,
No Definitions touch your senseless mind;
To you no Postulates prefer their claim,
No ardent Axioms your dull souls inflame;
For you, no Tangents touch, no Angles meet,
No Circles join in osculation sweet!

Mathematical satire retained its edge, and 'oscultation sweet' was displayed to great effect in James Gillray's 1792 picture of Pitt the Younger and the canvasser Albinia Hobart as 'A Sphere, projecting against a Plane'. The engraving came with mock 'Definitions from Euclid':

A Sphere, is a Figure bounded by a Convex surface; it is the most perfect of all forms; its Properties are generated from its Centre; and it possesses a larger Area than any other Figure....
A Plane, is a perfectly even & regular Surface, it is the most Simple of all Figures; it has neither the Properties of Length or of Breadth; and when applied ever so closely to a Sphere, can only touch its Superficies, without being able to enter it.

Another comic artist, Thomas Rowlandson, produced around the same time a series of illustrations showing different types of man on their characteristic types of horse. There was an 'astronomer'—on a horse that gazed heavenwards—and a surveyor who was escaping at the gallop from the scene of his crimes. And there was a 'calculator', a short-sighted gentleman earnestly scribbling in his book on the back of a notably slow horse, evidently oblivious to his surroundings. Possibly Rowlandson meant it for one of Maskelyne's 'calculators', or someone engaged in similar work. The caption suggested a connection with timekeeping:

> This is the finest Horse in the World for a calculator—keeps time to a Second! he goes tis true but like the finer wheels of a clock his motion is scarcely perceptible.

Evidently this calculator's work was absorbing, in the way mathematics was traditionally supposed to be. But, like Johnson's astronomer, his mind and his ability to function as a social being were not improved by the discipline and concentration his work required.

So as well as straightforward fun, mathematics could be the subject of serious anxiety. Rowlandson's calculator made himself a figure of fun by over-concentration on the subject. There was a hint in Domett's 'Ode' that being poor at mathematics might not be such a bad thing: that mathematical attainment, even at quite a modest level, might be unsuitable for a gentleman. Satire like that in 'The loves of the triangles' suggested that what had once been a model of clarity and precision was in danger of becoming an obfuscatory, dusty scholastic pursuit. Thus, at the same time that Hutton and others were arguing for the importance and value of mathematical education—the idea that it improved the mind was as frequently expressed as ever—others were troubled by its effects on the mind and on society.

The young Samuel Taylor Coleridge, writing in 1791, reckoned that mathematics tended to produce intellectual isolation and 'languidness' by starving the imagination while 'Reason is feasted', and produced a poem—'A mathematical problem'—intended to rescue the subject from that unhappy fate. It did so by rendering the statement and proof of a theorem in Euclid (to construct an equilateral triangle on a given line segment) into verse:

> From the centre A. at the distance A. B.
> Describe the circle B. C. D.
> At the distance B. A. from B. the centre

The round A. C. E. to describe boldly venture.
(Third postulate see.)
And from the point C.
In which the circles make a pother
Cutting and slashing one another,
Bid the straight lines a journeying go.
C. A. C. B. those lines will show.
To the points, which by A. B. are reckon'd,
And postulate the second
For Authority ye know.
A. B. C.
Triumphant shall be
An Equilateral Triangle,
Not Peter Pindar carp, nor Zoilus can wrangle.

Mathematics might have been a valuable study, but its effects could be questionable: precisely because it disciplined the mind and brought the imagination under control it was not to be embraced without question. Its promises of precision and certainty were not always a substitute for other qualities in the pursuit of excellence and virtue.

So, by the beginning of the nineteenth century, concerns about what mathematics might do to you were quite capable of overwhelming optimism about what it might do for you. We've seen, over the course of this book, a number of different ways in which eighteenth-century people might have been concerned about mathematics: sceptical of its achievements, nervous of its power, or doubtful of its benevolence. Early in this period it was still possibly for well-meaning commentators to believe that numerical or geometrical activity was essentially a form of conjuring: to dwell on its traditional associations with planetary influences and cosmic correspondences and hence to be deeply suspicious of it. Popular astrology, particularly in the form of the almanac but also of the private astrological practitioner like William Lilly, gave some justification to such concerns. Here, mathematical expertise really was at the service of practices which were at best an exploitation of some people's credulity, at worst a traffic with unorthodox powers.

The idea that mathematics was primarily a tool of deception—as a way for the powerful to keep themselves powerful and the rich to keep themselves rich, much like other forms of expertise perhaps—was something Poor Robin tapped into in his early satire, in which mathematics was frequently a metaphor for deception or dishonesty in personal or commercial life. It was reinforced and developed by mathematics' occasional public failures: astronomical events that didn't happen, commercial failures, and malice and incompetence in accounting. From the point of view of the ordinary people who ended up feeling hoodwinked, these incidents were further examples of the way mathematics could be used against you: the way in which it was a tool of oppression and deception.

At the same time, of course, that pessimism was always in tension with optimism about what mathematics could do for you. Many people were discovering that learning some mathematics was a fairly reliable way to get on in the world; some would go on to pursue spectacular careers in practical mathematics. Only a smattering of arithmetical competence was needed for everyday transactions, but many were the arithmetic primers which implicitly held out the promise of wealth and power to those who mastered their contents in full.

One result of that optimism was the founding of the Spitalfields Mathematical Society and other societies like it, where working men could meet, learn, puzzle, and instruct one another: each man with 'his pipe, his pot and his problem'. Another was the project pursued by the *Ladies' Diary* and other publications, to turn mathematics into a polite accomplishment. They promoted an idea that mathematics was not just a way to better yourself as a working man or woman—though it was that—but also a form of self-improvement at a more genteel level, a way to improve and beautify the mind.

In the *Diary* a code of modesty and decorum coexisted with the precision and detail needed to state and solve mathematical problems, producing a tension which was eventually resolved in favour of mathematical content and at the expense of inclusive politeness. The *Diary* became difficult, accessible to relatively few readers as a result. Mathematics lost its foothold in the world of polite accomplishment and polite discussion. Perhaps that was a fateful event in the long run, for mathematics and its reputation. In the shorter term one of the outcomes was a sense that too much mathematics was a bad thing both socially and personally: it would make you antisocial, and it would make you mad. Its promises of power and self-control were an illusion, and it would lead not to success but to delusion and despair. Hence Johnson's crazed astronomer, and hence Rowlandson's antisocial 'calculator'.

In a more public sphere, the credit of mathematics was still very high in the middle of the eighteenth century. Gentlemen's societies and popular lecturers preached a gospel of Newtonianism, whether complex and technical or easy and delightful, in which mathematics was ever-present as an idea, if not always as an object of detailed study. Many were the spheres in which mathematics helped to organize the mid-century world, from fenland drainage projects to the changing of the calendar. But a revolution in taste was at hand, and by the 1750s it was possible to wonder whether the line of beauty was always the natural curve, never the geometrical construction, and to point to contemporary trends in garden layout—for example—as evidence of a change in taste away from geometric design. Perhaps geometry was too cold, too inhuman, to be entrusted with the organization of the public world.

The final decades of the eighteenth century were nevertheless a period when the practical power of mathematics, and the degree to which it was embedded in certain parts of national life, were increasing. Through the military academies and specialist mathematical schools, British mathematical knowledge was disseminated to the battlefields of Europe and beyond, and through the study of navigation in those institutions it was carried across much of the globe. For the state, and for those involved in commerce, mathematics was the supreme instrument of information and control, and after

William Playfair's invention of graphs it took on a new lease of life in displaying and understanding the behaviour of all kinds of measurable quantities.

But still, all was not well. The dominance of natural history at the Royal Society, and the decline of British mathematical research since the time of Newton, led some to feel a sense of crisis by the late eighteenth century. Hutton's ill treatment by the Royal Society was the catalyst for a flurry of writings on the desirability of educational and institutional reforms which would favour mathematics. Those strident voices met stiff opposition, and several of the old anxieties about mathematics, its powers, and its status, came once more into view. Some seem to have felt that the demands of Hutton and others were excessive: that mathematics was important, but not *that* important. It was not, after all, a polite accomplishment, and if it was not a form of conjuring it was certainly a trade, and one whose effects on the mind were the subject of some misgivings. Was it, then, really a proper part of a gentleman's education?

So the scene was set for both the optimism and the pessimism of the nineteenth century regarding the power and the achievements of mathematics: for the objections of Dickens and his generation to a mentality of quantification, of 'facts, facts facts', but also for the promises that mathematics and mathematical education would present to new generations throughout the years to come.

> He form'd a line & a plummet
> To divide the Abyss beneath.
> He form'd a dividing rule:
> He formed scales to weigh;
> He formed massy weights;
> He formed a brazen quadrant;
> He formed golden compasses
> And began to explore the Abyss...

Figure 16 shows Urizen, from William Blake's complex personal mythology, worked out during the 1790s in a series of spectacularly

FIG 16 Urizen, William Blake, 1794
British Museum © Peter Willi/SuperStock

illustrated poems. This image was the frontispiece from his *Europe: A Prophecy* in 1794: it's probably one of the most iconic pictures of mathematics at work, and also one of the most ambiguous. For Blake the whole classical heritage, of which Euclidean geometry was a part, was suspect: 'Grecian is Mathematic Form: Gothic is Living Form', he wrote.

Urizen ('your reason') was one of the four 'Zoas' making up the fourfold individual for Blake: he stood for the rational faculty, but also for the individual's limits or horizons. He was responsible for a fallen world: the primeval priest, the archetypal king, and the deity and lawgiver. His activities were concerned in part with mathematical and geometrical law, with the wielding of his 'golden compasses'. Blake's well-known picture of Newton shows the great British mathematician in a strikingly similar pose (as does the massive statue inspired by it which now stands outside the British Library), adding a further layer to his ambivalence about the rule-bound world in which he found himself. Blake's exploration of Urizen thus became one of the high points of eighteenth-century anxiety about mathematical order and its relationship to the world and to human beings. Mathematics, here, bore fundamental importance, but brought inescapable dangers.

> Let each chuse one habitation:
> His ancient infinite mansion:
> One command, one joy, one desire,
> One curse, one weight, one measure
> One King, one God, one Law.

> The clock strikes, wide asunder start the gates, and in they come, a whole army of porters, darting hither and thither, and seizing the said bags, in many instances as big as themselves. Before we can well understand what is the matter, men and bags have alike vanished—the hall is clear... they will be dispersed through every city and town, and parish, and hamlet of England; the curate will be glancing over the pages of his little book to see what promotions have taken place in the church, and sigh as he thinks of rectories, and deaneries, and bishoprics; the sailor will be deep in the mysteries of tides and new moons that are learnedly expatiated upon in the pages of his; the believer in the stars will be finding new draughts made upon that Bank of Faith impossible to be broken or made bankrupt—his superstition, as he turns over the pages of his Moore—but we have let out our secret. Yes, they are all almanacks—those bags contained nothing but almanacks.

And what of Poor Robin? Bearer of both the promise of mathematics and its anxiety, articulator both of its usefulness and its ridiculous side, Robin had survived the years remarkably unscathed.

But after the death of Thomas Peat in 1780 *Old Poor Robin* never again seems to have found an author with a coherent vision for him.

There were increases in the quantity of mathematical and useful information, with the concomitant claim on the title page that the work was intended for the 'Entertainment and Improvement of the human Mind'. There were comic dialogues, songs, sometimes with printed music, scurrilous rhymes and slightly off-colour anecdotes. There was still much to laugh at in the world of the popular almanac, and Poor Robin continued to peddle much the same brand of wise-fool material that he had throughout the century. He remained a loyal whig with a quietist view of affairs. In 1793 he denounced Tom Paine and his radicalism at length. In 1797 he remarked apparently without irony that

> Honest, poor working men are the truest riches of every state . . . to seduce them with pretences of pointing out something more beneficial to them, to have and little to do, and thereby making them discontented and repining, is most hateful and abouminable.

But imagination was, possibly, wanting. In 1790 the author had been reduced to the dire expedient, stuck for a preface, of writing several pages about how hard it was to write a preface. During the 1790s he did away, at last, with the two calendars—serious and comic—and thus perhaps with the last remnant of what had once been distinctive about Poor Robin's almanac.

Around the turn of the century some energy was recaptured in Poor Robin's humour with a column of 'Observations' in which incidents from Robin's home life were recounted: his wife looking over his shoulder as he wrote and offering unhelpful suggestions, and so on. But from 1804, for about a decade, there was a deeply unfortunate interlude. A new compiler killed off the character of Old Poor Robin in a lengthy deathbed scene of extraordinary unpleasantness. The writing was supposedly taken over by his son, under whom broad, earthy, and downright gross humour increased markedly. The author responsible was sacked after the 1809 edition, but his replacement—despite a partial return to the old ways—had equally unfortunate tastes, this time for esoterica, and inserted into *Old Poor Robin* discussions of numerology and the cabbala. In 1816 the preface consisted of a discussion of the Universal Soul. He didn't lack for comic invention—in the same year Poor Robin took a trip to Jupiter by balloon—but he was obviously no more the right man for the job than his predecessor.

All of this was unpopular with customers. By the start of the nineteenth century *Old Poor Robin* had fallen behind Moore's and Wing's almanacs, the *Ladies Diary*, and two more, in popularity, and its profits were a somewhat marginal £8 a year (as against Moore's £2,500). Under the author of 1804–1809, sales fell from over 6,000 to around 3,000. *Old Poor Robin* never made a profit after 1811. This was, after all, the period of 'delicacy', which produced for example Bowdler's 'family' version of Shakespeare. Readers who had been loyal to the old Poor Robin apparently felt, in numbers, that his new manifestation had little to offer them.

In the 1820s *Old Poor Robin* made something of a return to form. The prefaces, which had been the vehicle for much of the problematic material over the previous few years, were dropped, and Poor Robin's exploits in the new decade included a jolly trip to China, together with whimsical descriptions of the science and astronomy he found there. Robin continued to advertise proudly the length of his heritage: the 1828 edition displayed on its title page that it was the 166th edition and that Poor Robin was still, as he always had been, a 'Well wisher to the Mathematicks'.

The contents now included the times of sunrise and sunset and the motions of the moon, together with columns of 'Pretty Poetry', 'Entertaining Rhymes', and 'Prose, Merry, Moral, and Miscellaneous'. There was no longer a mock calendar, but there were rambling reflections on fashionable life and political commentary:

> I wonder, then, what we're the better,
> If Whig or if Tory should reign?

When both of 'em, true to the letter,
Tax us over and over again.

There were wage tables, tide tables, a running description of Robin's troublesome home life, and even songs:

My wife was like a badger gray,
To turn her pate I bought her
A pot of bear's grease genuine,
Of Mister Gillingwater.

And Poor Robin still had his finger on the aspirations of mathematics, and its absurdities. The 1828 issue quoted an advertisement for

> J. & P. Prickalouse, Tailors and habit-makers to his Majesty King George, the Duke of Clarence, and all the Royal Family,
> By whom garments are mathematically cut and perfected upon the truest principles of equilibrium, and a thorough investigation into the philosophy of gravity, the curves of conic sections in the positions of the arm-pits, assisted by the nicest calculations of algebraical fractions to determine the fluctional quantities of length, breath, and circumstance; by which means, every piece of clothing made and perfected by Messrs. J. and P. P. fits inexpressibly.

For a few years it looked as though the situation had perhaps been rescued. But it was not to be.

The almanac business as a whole was thriving. At the start of the nineteenth century the Stationers' Company was printing twenty-five different almanacs every year and making a gross profit of nearly £4,000 by their sale. The profit was twice that in 1820, and remained steady up to about 1840. Nearly half a million almanacs circulated in most years: roughly one for every seven people in the country. A commentator a few decades later would place *Old Poor Robin*, with Moore and Partridge, 'in every cottage and workshop'. The description, quoted above, of 'Almanac day' at Stationer's Hall later in the nineteenth century gives a sense of the scale of the business.

Astrology may have become a stale trope in literature—'For Tycho and Copernicus agree, No golden planet bent its rays on me' (Mary Leapor)—and the influence of the heavens something slightly absurd: 'A hemisphere of evil planets reign! And every planet sheds contagious frenzy!' (Sheridan). But it retained its currency both in its judicial and its natural forms. As the poet Ann Thomas put it:

> Then they consult our conjurer too;
> Poor man—indeed he cannot see,
> But reads the stars like ABC;
> He tells them all what will betide,
> And when each lass shall be a bride;
> And when the destined youth appears,
> Describes the very coat he wears;
> He'd tell her too, if he may prove
> An object worthy of her love.

George Eliot would reflect on the situation in *Adam Bede*—set in 1799—through the character of Mr Craig the gardener, a man who 'knew his business' and indeed was astute enough to make some money from the garden as his own 'spekilation' although 'I've cot to run my calkilation fine, I can tell you, to make sure o' getting back the money as I pay the Squire.' Craig firmly believed in weather prognostication, although he was somewhat scornful of the almanac-makers: 'th' met'orological almanecks can learn me nothing, but there's a pretty sight o' things I could let *them* up to, if they'd just come to me.' As far as careful calculation was concerned, 'I should like to see some o' them fellows as make the almanecks looking as far before their noses as I've got to do every year as comes.'

All the same, some of their predictions did come true:

> Why, what could come truer nor that pictur o' the cock wi' the big spurs, an' th' firin', an' the ships behind? Why, that pictur was made afore Christmas, and yet it's come as true as th' Bible. Why, th' cock's France, an' th' anchor's Nelson—an' they told us that beforehand.

Bernard Capp, the historian of English almanacs, quotes a man 'high in office in the City of London' in this period who would 'sooner believe in Moore than in Bonaparte, or Mr. Addington'. Not all almanac readers were the ignorant and credulous, whatever their detractors may have said, and the quality of Moore's almanac in particular was high around 1800. The editor, Henry Andrews (a computer for the *Nautical Almanac*),

> was able to provide full explanations and descriptions of eclipses, comets and similar data, including accounts of the newly-discovered planet, Uranus, and the transit of Mercury in 1799. Like the best Stuart compilers, he read the *Transactions* of the Royal Society while writing for the masses.

For all that, the almanacs continued to contain both judicial and natural astrology and even prophecy, genres more controversial than ever. And some almanac authors could be shown not to believe in the predictions they compiled. Charles Hutton, the compiler of ten almanacs for the Stationers' Company, had sceptical words in his *Mathematical Recreations* for the traditional association of magic squares with astrology: 'those who can find any relation between the planets and such an arrangement of numbers, must no doubt have minds strongly tinctured with superstition'. And belief in judicial astrology was certainly very much on the wane, even if natural astrology remained considerably more respectable.

The combination of handsome profitability and high visibility with evident bad faith and the peddling of what some called 'superstition' made the sale of almanacs, despite their merits, a target for ever more hostile criticism, from those who believed the Stationers' Company was enriching itself by exploiting human weakness. The decisive blow was dealt in 1828 by Charles Knight, publisher to the recently formed Society for the Diffusion of Useful Knowledge and a campaigner for the improvement of the working-class mind (through, naturally, the purchase of Knight's books). The description of almanac day quoted above is, ironically, his. Charles II had seen Poor Robin's almanac in, his cousin's great-great-great-grandson, George IV, would see it out.

For Knight, the very processes of change that had taken place during the eighteenth century, including the education of more and more people in mathematical and scientific subjects, had made the almanac trade an embarrassing anachronism. He was offended by almanacs' repetition of aged, divisive political and religious slogans, by their discussion of starry influences for which neither science nor theology was prepared to vouch, and even by their moral tone.

He set out his case in (anonymous) articles in the *Athenaeum* and the *London Magazine*. He sincerely believed the Stationers' Company was failing in its duty to the newly literate by continuing to print such stuff. For him *Old Poor Robin* was 'a farrago of filth, obscenity, and stupidity', 'execrable poison', a representative of the insulting absurdities through which the Stationers' Company made money—lots of it—from a public taste for superstition and degraded buffoonery. The word *obscenometer* was coined for a device which would measure 'the degrees of indecency in printed matter'. Knight quoted from *Old Poor Robin*:

> If it don't snow
> I don't care.
> But if it freezes
> It may as it pleases
> And then I sneezes,
> And my nose blow.

He commented: 'Could any reader of this day imagine that in the year when the London University was opened, and the Society for the Diffusion of Useful Knowledge was beginning its work, he could find these lines at the head of the Calendar for January?'

Knight's arguments were based on overstatement and oversimplification. Many almanacs in reality contained no astrology at all, and were only purveyors of 'superstition' in the very general sense of weather predictions and the use of the lunar phases as a medical guide. In 1828 *Old Poor Robin* contained neither prognostication nor any innuendo more offensive than the word 'ass'—it's hard to see what harm his silliness was really doing to the public mind. Almanacs, as they always had done, presented a range of material to a range of readers, and Knight determinedly lumped it all together so as to condemn the whole business. He even argued, rather preposterously, that the officials of the Stationers' Company should be prosecuted under the 1824 Vagrancy Act, which forbade the sale of false prophecies. Prediction of the future should be a matter for science and statistics, not for astrologers.

In 1828 the Society for the Diffusion of Useful Knowledge launched a new and 'British' almanac in which superstition, crudity, and of course humour, had no place. The *British Almanac* contained no fiction, no poetry, and no puzzles, but plenty of statistics. It was not entirely new in this respect: the *Nautical Almanac* and a handful like it already contained only data. But for the makers of the *British Almanac* mathematics and science served an ideological purpose: precise, factual, uncontroversial, they would tend to unite and improve all classes of society, particularly the core audience of the intelligent working class.

The Stationers' Company responded to the *British Almanac* by a retreat. They toned down the remaining astrological content of their almanacs, removing some—though not all—of what might be offensive to the new world of useful knowledge and delicate taste. But *Old Poor Robin* couldn't be rescued. It had lost most of its following of readers, and it had been losing money for a decade and a half. The Stationers accepted the inevitable. The 1828 issue was the last.

The promoters of the *British Almanac* took the demise of Poor Robin as a victory, and there was some ugly crowing. A short-lived *Poor Humphrey's Almanac* appeared in 1829 to report Robin's death in disparaging terms:

> He expired... blaspheming and calling for brandy. His last words were a compound of drivelling idiocy, and heartless depravity.

Robert Chambers, another hostile critic, attributed the demise of Poor Robin, with its 'gross superstitions and even indecencies', directly to the appearance of the *British Almanac*, which he judged to have prompted 'a signal reform' in English almanacs. 'Since that period the publications of the Stationers' Company have kept pace with the growing requirements and improved tastes of the age.'

But *Old Poor Robin* had been ailing for some time, and it is hard to see how long it could have continued even without Knight's attacks. His end was less a triumph of reason than a commercial failure. Perhaps it was inevitable, in the conditions of the early nineteenth century, that the *British Almanac*'s statistics and facts should replace Poor Robin's drolleries, that its essentially urban outlook should replace the logic of moon and seasons of the old rural almanacs.

Symbolic of his decline was the fact that, from being represented as an active and mature man in the eighteenth century, the character of Poor Robin became elderly and troubled by the 1820s. This was the case even beyond the almanac itself: on Boxing Day in 1823 a pantomime was performed at Covent Garden entitled *Harlequin and Poor Robin*. Poor Robin, who ended up in the ducking stool, was portrayed as 'a ragged, frayed old man'.

It's difficult to feel that Poor Robin was doing anything wrong by this point, with his whimsies shorn of rudeness (almost) and his calendar devoid of astrology. The final issues even followed Moore's lead in taking note of the new astronomical discoveries of several asteroids and the planet Uranus. But he may have sensed that the writing was on the wall:

> Spare, oh spare! my invaluable Almanacks, those records of wit and wisdom... that succeeding generations may admire in them the purity of a language, once so generally spoken, but then extinct.

And afterwards?

A few years later the government abolished stamp duty on almanacs altogether. Prices more than halved, and the trade peaked for the Stationers' Company in 1837—the year in which a young Victoria unexpectedly became queen and empress—with sales of 600,000. But now rivals could compete with the Company on equal terms, and the result was a long decline for its almanacs. Partridge and the *Ladies' and Gentleman's Diary* fell by the way. The Stationers' Company attempted to clean up its image by launching a string of morally-improving titles: the *Evangelical Almanac* and the *Family and Parochial*, among others, and *Old Moore* eventually replaced the last of its astrology with popular science. Despite burgeoning sales of independent almanacs in the later nineteenth century (one peaked at over a million copies a year), just a century after the death of Poor Robin the Stationers' Company ceased publication of almanacs altogether.

Almanacs containing judicial astrology reappeared with tremendous success in *Zadkiel's Almanac* and others, from the 1830s. Charles Knight, in fact, ended up feeling his battle had been lost—the Society for the Diffusion of Useful Knowledge disbanded in 1846, although its *British Almanac* continued to 1914—but in a sense it had not. Judicial astrology may have persisted, as it does to this day, but natural astrology—the astrology of times and seasons—had gone, as had Poor Robin's satire upon it.

And the comic almanac tradition? It was far older than Poor Robin, and it lived on after him. Middle-class productions such as the *Pickwick Comic Almanac* and *Punch's Almanack*, both launched around 1840, owed a debt to Poor Robin. The *Comic Almanack*, with illustrations by George Cruickshank, acknowledged it by using Robin's subtitle: 'An Ephemeris both in Jest and Earnest'. A connection with the world of the eighteenth-century mathematical practitioners was maintained in the person of Olinthus Gregory, a Woolwich mathematical professor who edited *Old Moore* and was possibly involved with the production of some of these humorous almanacs. But Poor Robin's gently anarchic subversions, or indeed any sense that the author was a 'well-willer to the mathematics', were not in evidence.

A *Poor Rabbin's Ollminick* appeared in Belfast in 1861. It was written in dialect, and its purpose was not comic but to record that dialect for antiquarian purposes. Writes Maureen Perkins, historian of nineteenth-century almanacs, 'The memory of *Poor Robin* was alive, but somewhat altered in the recalling.'

Poor Robin's special humour of the wise fool who knows better than his betters did resurface, though, in a remarkable flourishing of almanacs written in regional dialect which took place in the north-east of England, in Yorkshire and Lancashire, from the 1840s.

This flourishing of creativity among self-taught manual workers generated such titles as the *Bairnsla Foaks' Annual*, the *Nidderdill Olminac* and the *Shevvild Chap's Almanac*. By 1877 there were about forty different titles and one—the *Original Halifax Illuminated Clock Almenack*—would continue until 1956.

These almanacs preserved both the form and the function of *Old Poor Robin*—some of them matched its physical form quite closely—celebrating local events and regional pride. Some, like the *Bairnsla Foaks' Annual*, also contained some mocking judicial astrology. The *Bairnsla*, known as 'the *Punch* of the North', came to resemble *Old Poor Robin* very markedly in some of its humorous predictions:

> Widows will roar for the loss of their husbands, owd Maids will sigh for't'want a wun.

Evidently Poor Robin's brand of humour retained its vitality even after his demise.

Poor Robin of Saffron Walden had enjoyed a remarkable, and a remarkably long, life. By the time of his death he was the sole survivor of a brand of astrological humour which had its origins in the political and social turmoil of 1660s London. He could look back over good King Charles' golden days, the fall of the house of Stuart and the House of Hanover's long occupation of the British throne.

Poor Robin's life was a period of social and intellectual change, and from the point of view of his own distinctive concerns—mathematics and why it was funny—it was a period that saw repeated swings of the pendulum, as different ideas about mathematics struggled for dominance. Was it absurd, this juggling with numbers and shapes? Was it presumptuous? Was it deadening? Was it, on the other hand, vital, powerful, and empowering? Was it a force for good in the hands of ordinary people? The debates would continue in the new worlds of the nineteenth and twentieth centuries, and into our own time. Poor Robin, with his finger on their pulse, had helped to shape them from the beginning. His story has made the perfect frame for the wonderful world of popular mathematics, throughout his long, long life.

ACKNOWLEDGEMENTS

This book relies heavily on the work of others. Many of them are named in the text, and it would be invidious to single out individuals for praise. An exception must be made, however, for E.G.R. Taylor, whose work on the mathematical practitioners of early modern Britain is without peer.

Particular thanks are due to All Souls College, Oxford, where I was a fellow during the period of research and writing for this book.

I am also very grateful to Latha Menon at Oxford University Press, who provided the stimulus for this book, Emma Marchant who guided it through the publication process, and Jacqueline Stedall, who read it in draft. Special thanks to Dan Harding for his work on the text in its final stages.

Most special thanks go to Jessica and William, who lived with Poor Robin for many months with great good humour and forbearance.

NOTES ON SOURCES

The purpose of these notes is to direct interested readers to those sources of important information and of substantial quotations which are not named in the main text. It is not an exhaustive list of sources used or consulted.

General

E.G.R. Taylor, *The Mathematical Practitioners of Hanoverian England, 1714–1840* (Cambridge, 1966) is invaluable as a source of reference for the whole period. For some more prominent individuals the information in the *Oxford Dictionary of National Biography* (*ODNB*, www.oxforddnb.com) is also valuable, while for the writing activity of various authors it has also been useful to consult the *English Short Title Catalogue* (estc.bl.uk), *Eighteenth-Century Collections Online* (galenet.galegroup.com/servlet/ECCO), and *Early English Books Online* (eebo.chadwyck.com).

Chapter 1

The main sources for this chapter are the almanacs themselves, and Poor Robin's other works including the *Weekly Intelligence*. Charles II's procession is described in John Ogilby, *The relation of his Majesties entertainment* (London, 1661); the quote from Samuel Hartlib is from his 1653 'Proposalls towards the advancement of learning', printed in Charles Webster, *Samuel Hartlib and the Advancement of Learning* (Cambridge, 1970). That from Margaret Cavendish is from her *Description of a new world, called the blazing-world* (London, 1666).

Bernard Capp, *Astrology and the Popular Press: English Almanacs, 1500–1800* (London, 1979), Cyprian Blagden, 'The Distribution of Almanacks in the Second Half of the Seventeenth Century', *Studies in Bibliography* 11 (1958), pp. 107–16, and Frank Palmieri, 'History, Nation and the Satirical Almanac, 1660–1760', *Criticism* (Summer 1998), are essential sources for the almanac and comic almanac tradition in the seventeenth and eighteenth centuries. My own 'Poor Robin and Merry Andrew: Mathematical Humour in Restoration England', *Bulletin of the British Society for the History of Mathematics* 22 (2007), pp. 151–9, is also possibly of use concerning the context of mathematical satire. The quotation from Maureen Perkins is from her *Visions of the Future: Almanacs, Time, and Cultural Change, 1775–1870* (Oxford, 1996);

I follow her analysis of the natural and judicial astrology traditions. Poor Robin as an author of pornography is suggested by Lois G. Schwoerer in the *ODNB* article on Henry Care. The only biography of William Winstanley is that in the *ODNB*. For William Lilly we also have Ann Geneva, *Astrology and the Seventeenth Century Mind: William Lilly and the Language of the Stars* (Manchester, 1995).

Chapter 2

The newspaper citations are chiefly from the *Post Boy* of 12 September 1699.

Sarah Dixon's poem was entitled 'Cloe to Aminta On the Loss of her Lover' and appeared in her *Poems on Several Occasions* (Canterbury, 1740). For the South Sea Bubble the key modern discussion is now Helen Paul, *The South Sea Bubble* (Abingdon, 2010). Paul is responsible for revising the traditional idea of a gambling fever and for characterizing the incident as a rational bubble. The older study by John Carswell, *The South Sea Bubble* (London, 1960; new edition Stroud, 2001) was also used for this chapter. The quotation concerning 'Devil take the hindmost' is from John Trenchard, 'A Letter of Thanks' (London, 1720). The contemporary quoted concerning John Grigsby was the anonymous author of 'The secret history of the South-Sea scheme', printed in *A collection of several pieces of Mr. John Toland* (London, 1726).

The quotation from Nicholas Rodger is from *The Command of the Ocean: A Naval History of Britain, 1649–1815* (London, 2004); the sailor's complaint appears in a broadside of 1700 entitled 'An account of the many frauds and abuses, which have been frequently committed... to the great prejudice, and discouragement of Sea-men'. The 'dangerous mathematics' is from 'Captain Gulliver' [i.e. Eliza Heywood?], *Memoirs of the court of Lilliput. Written by Captain Gulliver...* (London, 1727); the marooned philosophers and geometricians appear in M. l'abbé Desfontaines (Pierre-François Guyot), trans. J. Lockman, *The travels of Mr. John Gulliver, son to Capt. Lemuel Gulliver...* (London, 1731).

Chapter 3

The exercise books of Isaac Hatch, Robert Gardner, and Ann Mohun are now in a private collection and have not previously been studied. Biographical information on their writers is derived from the International Genealogical Index (IGI, www.familysearch.org). For Hatch this is supplemented by the will of Elizabeth Hatch, his mother, which is preserved with the exercise book.

A useful source for mathematics education in this period and beyond is Geoffrey Howson, *A History of Mathematics Education in England* (Cambridge, 1982); another is John Denniss, 'Learning Arithmetic: Textbooks and Their Users in England 1500–1900', in *The Oxford Handbook of the History of Mathematics*, edited by Eleanor Robson and Jacqueline Stedall (Oxford, 2009), pp. 448–67. The difficult subject of 'mathematics avoidance' is addressed by Jane Wess in 'Avoiding Arithmetic, or the

Material Culture of Not Learning Mathematics', *Bulletin of the British Society for History of Mathematics* (2012).

The records of the Emery school are held by the Bedfordshire and Luton Archives and Records Service. I am grateful to James Collett-White for his assistance with this material. Thomas Salmon is exhaustively discussed in Benjamin Wardhaugh, *Thomas Salmon: Writings on Music* (Aldershot, 2013).

John Aubrey's manuscript list of mathematical instruments (Bodleian Library, MS Aubrey 10, fol. 109r) is online as part of the Bodleian library's exhibition 'John Aubrey and the Development of Experimental Science': www.bodleian.ox.ac.uk/about/exhibitions/online/aubrey. Information about the various charity schools named in the text is taken from descriptions of the schools or collections of their rules printed during the eighteenth century.

The quotation from Ruth Wallis is from her article on Walkingame in the ODNB.

The history of the Spitalfields Mathematical Society is described in J.W.S. Cassels, 'The Spitalfields Mathematical Society', *Bulletin of the London Mathematical Society* 11 (1979), pp. 241–58; as explained there, the original sources seem to have vanished during the Second World War.

Chapter 4

The Lowther/Frank accounts book is online at the Perdita Project: www.amdigital.co.uk/collections/Perdita.aspx. Only quite minimal biographical information is available for any of the women who wrote in the book, but something can be gleaned from the article about Sir John Lowther of Lowther in the *ODNB*, as well as from the IGI and the book itself. Concerning women's account books, a useful source is Judith Spicksley, 'Two Seventeenth-century Female "Accountants": Joyce Jeffreys and Sarah Fell', *Bulletin of the British Society for History of Mathematics* 6 (2005), pp. 1–8.

The copy of Smart's *Tables* discussed is in a private collection. The material concerning William Hunt is from his books *A Guide for the Practical Gauger* (London, 1673) and *The Gaugers Magazine* (London, 1687).

John Thompson's plane table and his telescopic level are in the Museum of the History of Science in Oxford: www.mhs.ox.ac.uk; information about him and his instruments can be found on the museum's website. The Adams firm is exhaustively documented in John Millburn, *Adams of Fleet Street: Instrument Makers to King George III* (Aldershot, 2000); further information is in John R. Millburn, 'The Office of Ordnance and the Instrument-making Trade in the Mid-eighteenth Century', *Annals of Science* 45 (1988), pp. 221–93; the bill quoted is taken from the latter.

The calculating machines of Samuel Morland and others are discussed in E.G.R. Taylor, *The Mathematical Practitioners of Tudor & Stuart England* (Cambridge, 1968). The almanacs of James and Benjamin Franklin are discussed in Perkins, *Visions of the Future.*

Chapter 5

Thomas Porcher's exercise book is in a private collection, and information about the family is recorded in an associated manuscript, supplemented from the IGI. Samuel Davis' interleaved copy of Brown's *Problems in Practical Geometry* is in a private collection.

The story of 'Automathes' appears in John Kirkby, *The Capacity and Extent of the Human Understanding; Exemplified in the Extraordinary Case of Automathes* (London, 1745).

The quotes from John Arbuthnot are from his *Essay on the Usefulness of Mathematical Learning* (Oxford, 1701); that from Byron appeared in 'Conversations of Lord Byron, by the Countess of Blessington', in the *Gentlemen's Magazine*, April, 1834. The story about Hobbes is taken from John Aubrey's *Brief Lives* (Oxford, 1898); the death of Archimedes is quoted from John Dryden's translation of Plutarch's *Lives* (1683).

For the *Ladies' Diary* the main sources are the works of Shelly Costa: 'The "Ladies' Diary": Gender, Mathematics, and Civil Society in Early-Eighteenth-century England', *Osiris* (2nd series), 17 (2002), pp. 49–73 and 'The Ladies' Diary: Society, Gender and Mathematics in England, 1704–1754' (Ph.D. Dissertation, Cornell University, 2000). Also of value are Teri Perl, 'The Ladies' Diary or Woman's Almanack, 1704–1841', *Historia Mathematica* 6 (1979), pp. 36–53; and Joe Albree and Scott H. Brown, '"A Valuable Monument of Mathematical Genius": *The Ladies' Diary* (1704–1840)', *Historia Mathematica* 36 (2009), pp. 10–47.

For Thomas Peat the sources are Taylor's *Practitioners* and the *ODNB*, as well as his own works and the issues of *Poor Robin* and *Old Poor Robin* he compiled. The history of the Stationers' Company in this period is discussed in Robin Myers, *The Stationers' Company: A History of the Later Years* 1800–2000 (Chichester, 2001) and Cyprian Blagden, *The Stationers' Company: A History*, 1403–1959 (London, 1960), as well as by Maureen Perkins and Bernard Capp.

Mathematics and madness is addressed in Alice Jenkins, 'Mathematics and Mental Illness in Early Nineteenth-century England', *Bulletin of the British Society for the History of Mathematics* 25 (2010), pp. 92–103.

Chapter 6

Much valuable information about the Spalding Gentlemen's Society and its members and correspondents appears in Diana and Michael Honeybone (eds), *The Correspondence of the Spalding Gentlemen's Society*, 1710–1761 (Woodbridge, 2010). John Muller is also discussed in W. Johnson, 'The Woolwich Professors of Mathematics, 1741–1900', *Journal of Mechanical Working Technology* 18 (1989), pp. 145–94. Edmund Scarburgh's Euclid was published as *The English Euclide* (Oxford, 1705).

For John Theophilus Desaguliers there is now a full-length biography: Audrey T. Carpenter, *John Theophilus Desaguliers: A Natural Philosopher, Engineer and Freemason in Newtonian England* (London, 2011). The *ODNB* has a valuable article on the Rainbow Coffee House group. Jeffrey Wigelsworth discusses Desaguliers at some length in his *Selling Science in the Age of Newton: Advertising and the Commoditization of Knowledge* (Farnham, 2010). 'All Hail Philosophy' is from Benjamin Martin, *The Young Gentleman and Lady's Philosophy* (London, 1759).

The change of calendar in 1752 is discussed in Elizabeth Jane Wall Hinds, 'Sari, Sorry, and the Vortex of History: Calendar Reform, Anachronism, and Language Change in *Mason & Dixon*', *American Literary History* 12 (2000), pp. 187–215 and Robert Poole, '"Give Us Our Eleven Days!": Calendar Reform in Eighteenth-century England', *Past & Present* 149 (1995), pp. 95–139. Newton's unpublished proposals are discussed in Ari Belenkiy and Eduardo Vila Echagüe, 'History of One Defeat: Reform of the Julian Calendar as Envisaged by Isaac Newton', *Notes and Records of the Royal Society of London* 59 (2005), pp. 223–54. The lines from Elizabeth Tollet are taken from 'On the Prospect From Westminster Bridge'; those from Susanna Blamire appear in 'Stoklewath; or, the Cumbrian Village'.

Richard Dunthorne and the work of the calculators for the *Nautical Almanac* are described in Mary Croarken, 'Providing Longitude for All: The Eighteenth Century Computers of the *Nautical Almanac*', *Journal of Maritime Research* (October 2002) and in David Alan Grier, *When Computers were Human* (Princeton, 2005). Croarken's article 'Astronomical Labourers: Maskelyne's Assistants at the Royal Observatory, Greenwich, 1765–1811', *Notes and Records of the Royal Society of London* 57 (2003), pp. 285–98 is also valuable in connection with this particular mathematical network. The quotation from Nicholas Rodger is from *The Command of the Ocean*.

Chapter 7

Robert Sandham's letters are quoted in O.F.G. Hogg, *The Royal Arsenal: Its Background, Origin, and Subsequent History* (London, 1963). Further information on the Royal Military Academy, and on Charles Hutton, is taken from Johnson, 'The Woolwich Professors' and the same author's 'Charles Hutton, 1737–1823: The Prototypical Woolwich Professor of Mathematics', *Journal of Mechanical Working Technology* 18 (1989), pp. 195–230. Some light is shed on Hutton's almanac-writing activities by both Perkins, *Visions of the Future* and Cyprian Blagden, 'Thomas Carnan and the Almanack Monopoly', *Studies in Bibliography* 14 (1961), pp. 23–44. Patrick O'Brian's biography, *Joseph Banks: A Life* (London, 1987), is valuable for the information about the incident at the Royal Society in 1783–4.

The illustrated gunner's perpendicular is in the Museum of the History of Science at Oxford, and once again information about it is taken from the museum's website, www.mhs.ox.ac.uk. The eighteenth-century changes in ballistics are discussed in W. Johnson, 'Benjamin Robins' *New Principles of Gunnery*', *International Journal of Impact Engineering* 4 (1986), pp. 205–19 and Brett D. Steele, 'Muskets and Pendulums: Benjamin Robins, Leonhard Euler, and the Ballistics Revolution', *Technology and Culture* 35 (1994), pp. 348–82.

Richard Shittler's copy of *The Young Man's Book of Knowledge* is in a private collection. It has previously been discussed in my *How to Read Historical Mathematics* (Princeton, 2010).

Chapter 8

My discussion of a 'culture of luck' depends in part on valuable unpublished work on the English lotteries by Natasha Glaisyer. The 'scheme' of 1730 was entitled *A Scheme for a New Lottery for the Ladies* (London, 1730). The 'justly enraged mob' appear in Richard King, *The Cheats of London Exposed* (London, 1780); Fielding's song in his *The Lottery* of 1732. John Molesworth's words are quoted from his *Proofs of the Reality and Truth of Lottery Calculations* (London, 1774), and Samuel Clark's refutation from his *Considerations Upon Lottery Schemes in General* (London, 1775). 'Review the Scheme...' is from a broadside, 'Ways & Means', produced by T. Bish in about 1781.

An exhaustive chronology of recreational mathematics appears at www.eldar.org/~problemi/singmast/recchron.html, by David Singmaster; some samples appear in my *A Wealth of Numbers* (Princeton, 2012). The modern translation of the Greek Anthology quoted is that of W.R. Paton (1918).

Mathematical poems are quoted from the *Poems* of Thomas Campbell, the *Poems* (London, 1833) of Alfred Domett, and the *Poetical Works* (London, 1834) of Samuel Taylor Coleridge. 'The loves of the triangles' appeared in *The anti-Jacobin* in 1798 and also in Frere's *Works* (London, 1872). I owe the characterization of the class divide between practical and gentlemanly mathematics to unpublished work by Shelly Costa. The quotations from William Blake are from *The [First] Book of Urizen* (Lambeth, 1794).

The description of 'almanac day' appeared in Knight's *Cyclopaedia of London* (1851) and is quoted in Myers, *The Stationers' Company*. Some quotations from the (very rare) late issues of *Old Poor Robin*, and some information about his trajectory in the early nineteenth century are taken from Perkins, *Visions of the Future*, but once again a main source is the almanacs themselves. Sales figures and other information about the Stationers' Company in this period are from Myers, *The Stationers' Company*, Blagden, 'Thomas Carnan and the Almanack Monopoly', and Blagden, *The Stationers' Company*; quotations from Capp are from his *Astrology and the Popular Press*.

Ann Thomas' lines appear in 'To Laura, On the French Fleet parading before Plymouth in August 1779'. The discussion of the demise of Poor Robin is based on—as well as the contemporary sources—Perkins, Capp, and Blagden; that of the dialect almanacs on Perkins alone.

INDEX

1/2017 Skoob